Magical Elements of The Periodic Table
Presented Alphabetically By
The Elemental
D Ra G O N S
In
Iker
The Indium Dragon
Pt
Paedra
The Platinum Dragon
Fr
Francine
The Francium Dragon
Ti
Tilly
The Titanium Dragon
Mo
Maximo
The Molybdenum Dragon
D20
Dusa
The Deuterium Dragon
C
Cole
The Carbon Dragon
Zr
Zora
The Zirconium Dragon
Bi
Bitsy
The Bismuth Dragon
By Sybrina Durant with Illustrations by Pranavva et al

Magical Elements of the Periodic Table

Presented Alphabetically By

The Elemental

Soft Cover Print ISBN– 978-1-942740-41-4

BISAC Codes:

JUV002270 JUVENILE FICTION / Animals / Dragons, Unicorns & Mythical

JNF016000 JUVENILE NONFICTION / Curiosities & Wonders

JNF051080 JUVENILE NONFICTION / Science & Nature / Earth Sciences / General

Hard Cover Print ISBN 13 - 978-1-942740-43-8 - ISBN 10—1-942740-43-3

EPUB ISBN 13 - 978-1-942740-40-7 - ISBN 10 - 1-942740-40-9

Contact Sybrina@sybrina.com.

Elemental Dragons
present
"Magical Elements Of The Periodic Table."

In this unique alphabet book, members of the Elemental Dragon Clan present 23 Magical Pure Elements of the Periodic Table plus 1 isotope-Deuterium, 1 compound-Quicklime, and 1 alloy-White Gold in alphabetical order. Each dragon has a magical tail tipped with an element which gives them unique powers.

Antz starts out the book by introducing the very necessary element Antimony on his element page.

Zora rounds out the alphabet by presenting facts and other fun information about the metal, Zirconium on her element page.

There is one dragon in the book whose magical tail is tipped with an alloy created by combining 5 different elements.

That would be, White Wing, the dragon with the White Gold tipped tail. He will reveal some interesting things about that metal alloy.

Dusa, the dragon with the Deuterium tipped tail will help readers understand what an isotope is and which pure element it comes from.

While not all elements in the Periodic Table are represented by letters of the alphabet, some in this book are introduced by alternate designations. For instance, the letter V is represented by the dragon with the Vital Calcium Tipped Tail.

The Elemental Dragon Clan and their techno-magical unicorn friends (from Book 1) are the perfect group to introduce you to metals and other elements in the Periodic Table. Hopefully, the Magical Elements of this periodic table book will spark an interest in the magical and real world properties of all the metals and other elements known today. You may be surprised at how prominently they feature in our every day lives.

Each element page in this book contains terms that might not be completely familiar to the reader. Refer to the definitions in the back of the book to get a clear understanding of each meaning.

There is also a fun elemental themed Periodic Table at the back of the book. It features metal horn unicorns, dragons with element tipped tails, wizards and knights with elemental staffs and swords; and radiated goblins.

Remember, "No metal - No Magic. . .and No Technology".

It's Techo-Magical.

Note: The reference page for all entries in this periodic table book is at MagicalPTElements.com/MDAPT. Follow sybrinapublishing on Instagram, Magical Elements of the Periodic Table on Facebook, @sybrinad on Pinterest, Sybrina_SPT on Twitter; and Sybrina Durant on LinkedIn.

A Word About

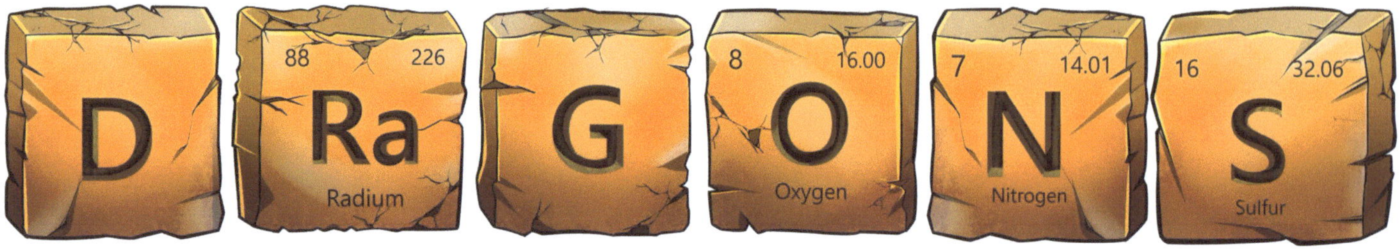

Dragons have long captured the our imaginations and often depicted as creatures of immense power and mystery. One compelling way to flesh out the lore surrounding these majestic beings is by linking them to elements from the periodic table, particularly focusing on those that may lend themselves to fantastical traits—like Elemental Dragons with their magical metal tails.

In this book, you will meet several elemental dragons. In the realm of transition metals, we find the Carbon Dragon. His shiny black and gray scales sparkle brightly. This wise dragon can conduct electricity and draw power from the air around him. Using the tip of his tail like a magic wand, Cole can create spectacular electrical storms, lighting up the sky like lightning in the night.

The Hydrogen powered Dragon is small and quick, covered in bright green scales. Hildy is full of energy and fun. With just a flick of her tail, she can create powerful bursts of energy, causing flames to dance in the air around her. Her light and graceful movements show how important hydrogen is in making water and its role in the universe.

The Lithium Dragon has beautiful lavender wings, representing peace and the ability to create energy. Lillian can heal emotional wounds and help balance troubled minds, making her a guardian of calm. This reflects lithium's importance in helping people with mental health through medicine.

The Oxygen Dragon is a lively creature with shiny scales that reflect many colors, like a rainbow shining over green fields. Ozzy loves to have fun and chooses joy over seriousness. With his amazing power, he can make any living thing grow ten times larger in just a moment. Like oxygen, which is vital for life, the Oxygen Dragon brings happiness and amazement to everyone who sees him.

The Platinum Dragon shines like the sun and represents wealth and success. Paedra has the magical ability to change regular materials into valuable platinum. As a symbol of luck and purity, the Platinum Dragon governs chance and fortune.

As we go further into this exciting world, we encounter the Sodium Dragon with its lovely pale pink scales that shine with energy. Known for being a bit unpredictable, Sorn can create huge waves of energy. His special power lets him control water, making whirlpools and strong currents, showing that he is powerful in his element.

The Titanium Dragon is a strong figure with tough, turquoise colored scales. Known as the Guardian of Strength, Tilly can call upon the earth's minerals to make her defenses stronger and show her great power in battles. She can even change the ground beneath her feet, creating shields or causing shakes in the earth, highlighting titanium's role in building and manufacturing.

In this creative mix of science and fantasy, dragons with magical metal tails are not just amazing creatures. They represent important elements and help create a rich story that connects the mysterious qualities of dragons with the basic parts of our world. They embody the essence of creation and change, and they represent the powerful forces that shape our universe. Through them, along with all the others in this book, you will to discover the magic found within the periodic table, unveiling a world full of wonder and connections between nature and fantasy.

MEET THE CLAN

THE DRAGONS WITH THE MAGICAL ELEMENTAL TIPPED TAILS

Antz Presents Antimony

Symbol: Sb Atomic Number: 51 Atomic Mass: 121.760

Antimony Facts

- In use since 4000-5000 BCE
- Brittle, crystalline solid
- Weak conductor of heat and electricity
- Strongly antiferromagnetic
- Neither ductile or malleable
- Metalloid

Antimony naturally occurs in the mineral stibnite. The largest crystals of stibnite are mined in Japan, China and the Baia Mare region of Romania.

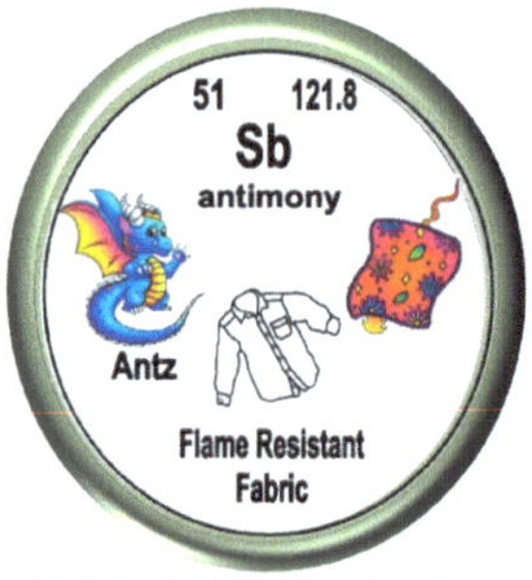

Antz's Magical Abilities

- Antz can wave his Antimony tipped tail over a flame and the fire will die out.

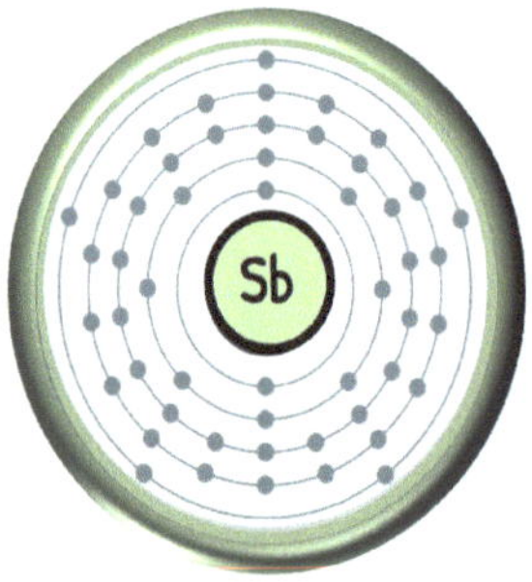

Atomic Structure

Uses For Antimony

Antimony is alloyed with tin to make the utensils we use in our kitchens!

Antimony alloys are used in making typefaces for clear, sharp printing.

Antimony(III) oxide is added to some glass, like television screens, to remove bubbles .

Antimony trioxide compound is primarily used in flame-retardant formulations.

Compounds of antimony are also used to make paints, glass, pottery and ceramics.

Did You Know?

- Cleopatra used stibnite (the source of Antimony) to make her mascara. She would crush the rock into a black powder and put it on her eyes to make them look pretty. This mascara was called "khol."
- In olden times, people used a pill made from Antimony to help with tummy troubles. It made them go to the bathroom. Here's the yucky part. They would save the pill to use again later. Nowadays, we don't do that because we now know it's not clean or good for our health.
- Mozart believed that Antimony was the top of the line medication of his time for fevers and other sickness. Some people think that the famous composer's doctor gave him too much and it became poisonous for his body and caused him to die.

Bitsy Presents Bismuth

Symbol: Bi Atomic Number: 83 Atomic Mass: 208.98

Bismuth Facts

- Discovered in 1753 by Claude Geoffrey Junine
- Solid
- Poor conductor of electricity
- The most diamagnetic metal
- Some ductility and alloys are malleable
- Post Transition Metal

Bismuth naturally occurs in the Earth's crust as Bismuthimite and Bismite ores. The largest Bismuth production is located in Vietnam and China. Also found in Bolivia, Peru, Mexico and Canada.

Bitsy's Magical Abilities

- When Bitsy taps her Bismuth tipped tail on another's belly, it rids them of indigestion.

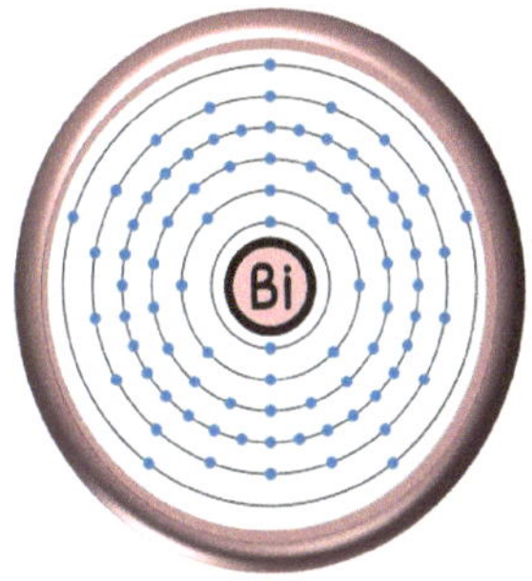

Atomic Structure

Uses For Bismuth

Basic Bismuth carbonate is taken in tablet or liquid form to soothe stomach aches from indigestion.

It is used as a catalyst in rubber production for things like gloves, boots, buckets and even rubber duckies.

Bismuth is used as a replacement for Lead in bullets and fishing weights.

When alloyed with Tin or Cadmium, Bismuth is used in fire detectors, extinguishers, electric fuses, and solders.

Did You Know?

- Bismuth metal does not rust when exposed to water and oxygen.
- If you take too much medication containing bismuth, you might get a black line on your gums next to your teeth. It's a condition called Bismuthia. Luckily, this is very rare.
- When liquid bismuth gets really cold, it doesn't get smaller like most things do. Instead, it actually gets bigger! This happens because it turns into ice-like crystals. There are four more elements that also do this: silicon, gallium, antimony, and germanium.
- Bismuth originates from the Latin word "bisemutum" and the German word "wissmuth", which means white mass.

Cole Presents Carbon

Symbol: C Atomic Number: 6 Atomic Mass: 12.01

Carbon Facts

- Element first proposed in 1789
- Soft, solid
- Strong conductor of heat and electricity
- Has a net magnetic moment of zero
- Neither ductile or malleable
- Non-Metal

The largest deposits of Carbon are found in the deep ocean, atmosphere, and sediments. It naturally occurs as coal, diamond, and graphite. Diamonds formed in the earth's mantle rise to the surface in kimberlite pipes.

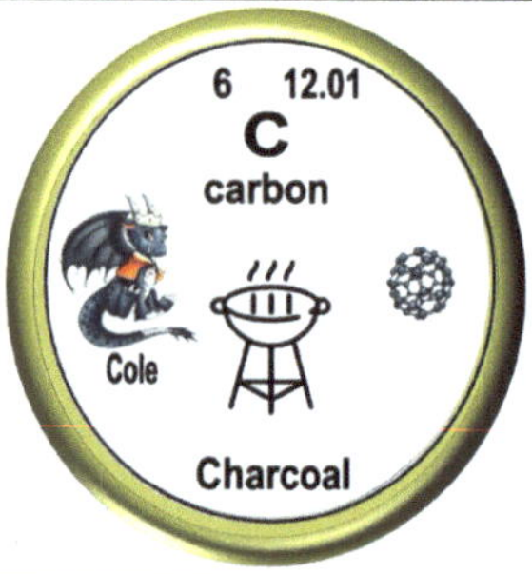

Cole's Magical Abilities

- With the touch of Cole's carbon tipped tail, he can purify air and water.

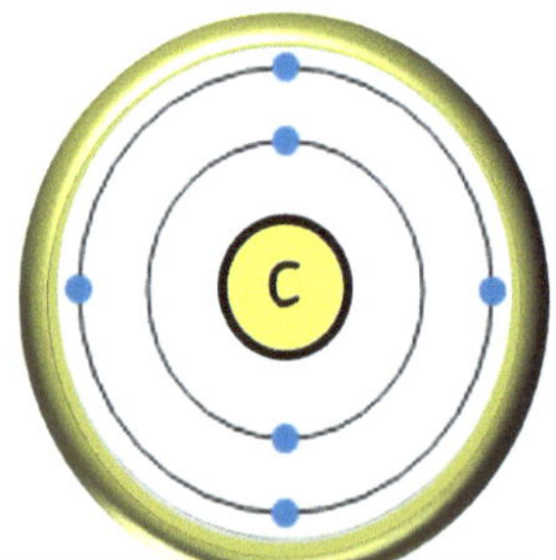

Atomic Structure

Uses For Carbon

Coal is used as a key energy source in cement production for bridges and buildings.

Carbon fiber is used to make airplanes and rockets.

Industrial diamonds are used for drilling and cutting rocks.

Carbon dioxide is essential for plant's survival.

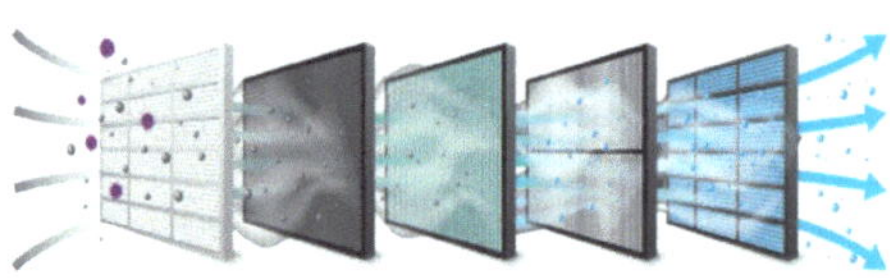

Graphene can be used as an ultrafiltration medium to destroy airborne viruses.

Pencil leads are made from carbon graphite.

Did You Know?

- Carbon has 4 natural allotropes. They are coal, graphite, diamonds and graphene. Graphene is the thinnest, lightest, strongest, most stretchy material ever created. Allotropes are naturally formed through nature. The only other elements that form allotropes are tin, sulfur, phosphorus, and oxygen.
- C60 Buckminsterfullerenes or Buckyballs were discovered in 1985. They have a cage-like fused-ring structure made of twenty hexagons and twelve pentagons, and resembles a soccer ball. They are considered the universe's weirdest molecule. They could one day replace silicon – the element on which computers and other electronic devices depend.
- Because Carbon forms more compounds than any other element, it is sometimes called the "King of the Elements."

Dusa Presents Deuterium (Hydrogen Isotope)

Symbol: D (or 2H) Atomic Number: 1 Atomic Mass: 2

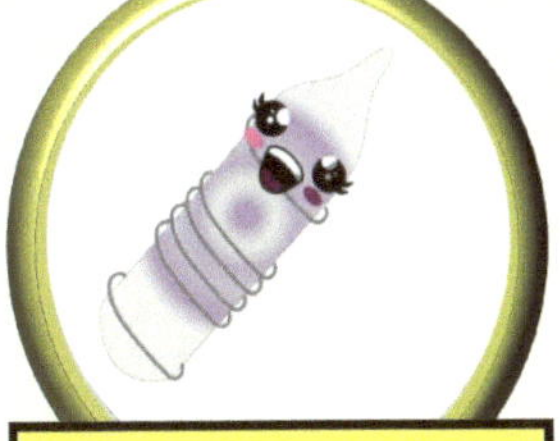

Deuterium Facts

- Deuterium Isotope discovered in 1931– Harold Urey
- Electrical conducting is characteristic of liquid metal
- Has a freely diffusible nuclear magnetic resonance
- Colorless, odorless gas that is easily ignited.

Deuterium is also known as Hydrogen-2 or Heavy Hydrogen. It is a naturally occurring stable isotope and can be readily extracted from the oceans. The largest production facilities are in the Canada and USA.

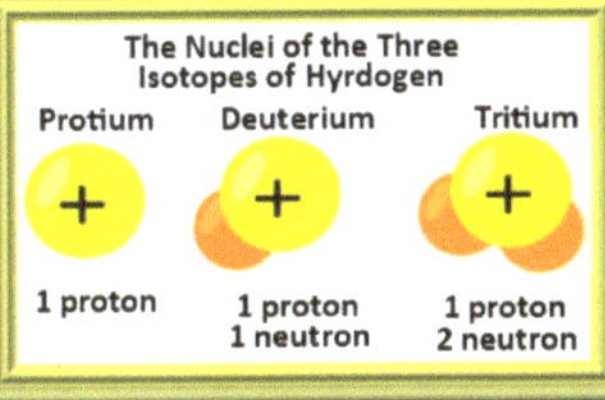

Dusa's Magical Abilities

- A touch of the deuterated tip of Dusa's tail cures psoriatic arthritis and other ailments.

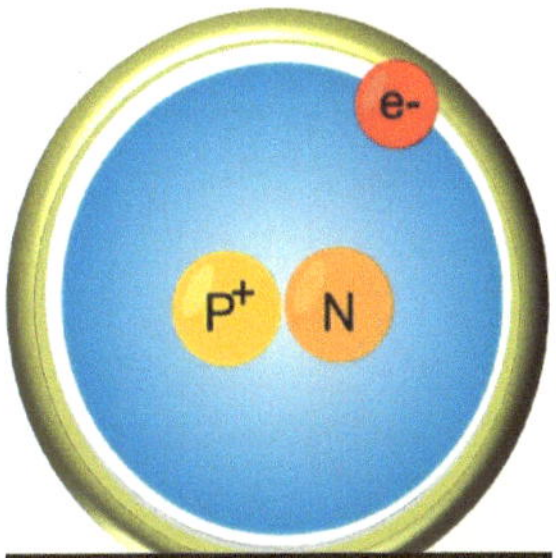

Atomic Structure

Uses For Deuterium

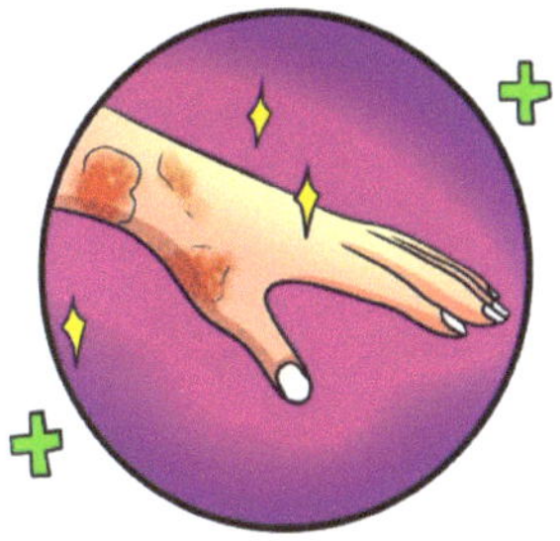

Deuterated drugs are being used for physical ailments such as Psoriatic Arthritis, Lupus and Crohn's disease

The current best bet for future fusion power plants is deuterium-tritium fuel.

By using Deuterated compounds, manufacturers can produce better-performing electronics, faster microprocessors, and see an increased life span of their devices, including OLED screens.

Did You Know?

- Nearly all Deuterium is thought to be left over from the Big Bang. It is not radioactive.
- The fusion energy released from just 1 gram of deuterium-tritium fuel equals the energy from about 2400 gallons of oil.
- Deuterated water (HDO) occurs naturally in normal water and can be separated through distillation, electrolysis, or chemical exchange processes. This "heavy water" is 10.6% denser than normal water and it does not have the blue color of regular water. Heavy water can be toxic to humans, but a large amount would be needed for poisoning to occur.
- An ice cube made of deuterium oxide will sink in normal water.

Euell Presents Europium

Symbol: Eu Atomic Number: 63 Atomic Mass: 152.0

Europium Facts

- Discovered in 1901 Eugène-Anatole Demarcay
- Soft, solid metal
- Electrical conductor
- Both paramagnetic and antiferromagnetic
- Ductible and malleable
- Lanthanide Rare Earth Metal

Europium is found in Monazite and other ores. Large deposits are located in India, Madagascar, and South Africa. It first was separated from the mineral samaria in magnesium-samarium nitrate in 1896.

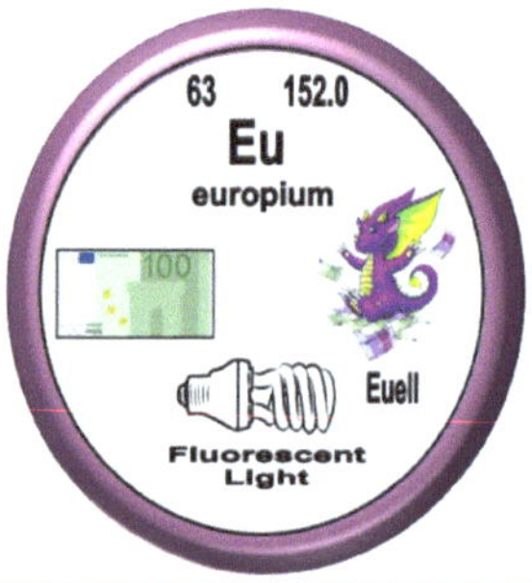

Euell's Magical Abilities

- Euell has the magical ability to turn any paper into money.

Atomic Structure

Uses For Europium

Europium is used in making low-energy light bulbs to produce soft light like those in incandescent bulbs.

Europium is used to print Euro bank notes. It is also used in phosphors in anti-forgery marks on Euro bank notes. Europium will usually have a red glow under UV light.

Europium is a source of blue light in light-emitting diodes (LEDs).

Europium compounds are used to produce quantum memory chips. The unit for storing information in quantum chips is "qubits".

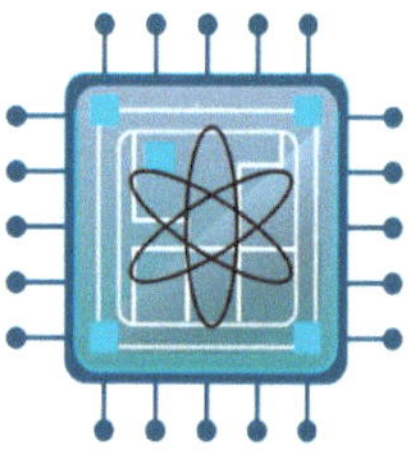

Europium(III) oxide is used to make special covers for greenhouses that help plants absorb more red light, which they really like.

Did You Know?

- Europium is distinct for being the most reactive, the rarest, the least dense, and the softest rare-earth element. It is so reactive that it must be stored beneath a non-reactive fluid to protect it from atmospheric oxygen or moisture.
- Europium was named after the continent of Europe.
- Europium is used in Euro bank notes to stop people from making fake money. The special ink with Europium glows red when you shine ultraviolet (or black) light on it, making it easier to catch anyone trying to print up money with the wrong ink. Even though there are other things that can glow, Europium is most like used on Euros because its name has the word Europe in it and it glows.

Francine Presents Francium

Symbol: Fr Atomic Number: 87 Atomic Mass: 223

Francium Facts

- Discovered in 1939 by Marguerite Perey.
- Presumed to be metallic
- Only has 22 min. half-life.
- Weak conductor of heat and electricity
- Is both ductile and malleable
- Radioactive Alkali Metal

Francium is extracted from Uranite. There is estimated to be one Francium atom for every one billion-billion atoms of Uranium. It was discovered while analyzing Actinium's decay sequence.

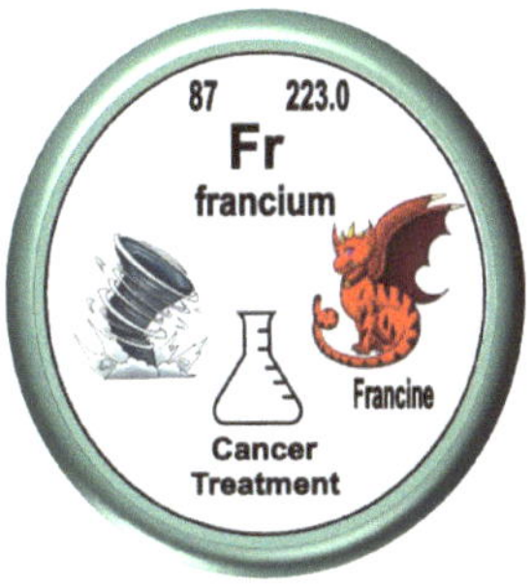

Francine's Magical Abilities

- Francine can cause a typhoon or cyclone with a wave of her Francium tipped tail over a body of water.

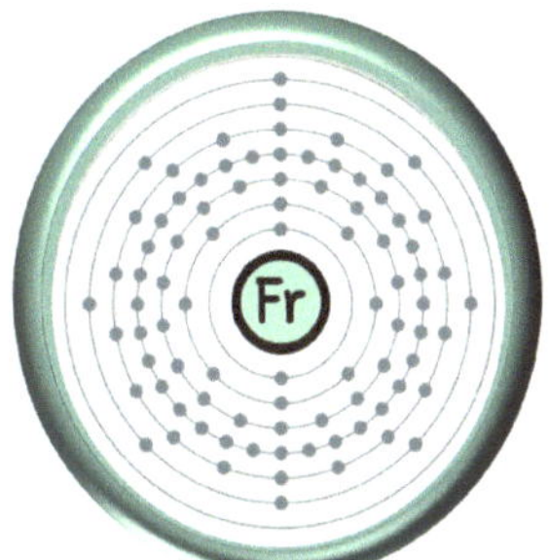

Atomic Structure

Uses For Francium

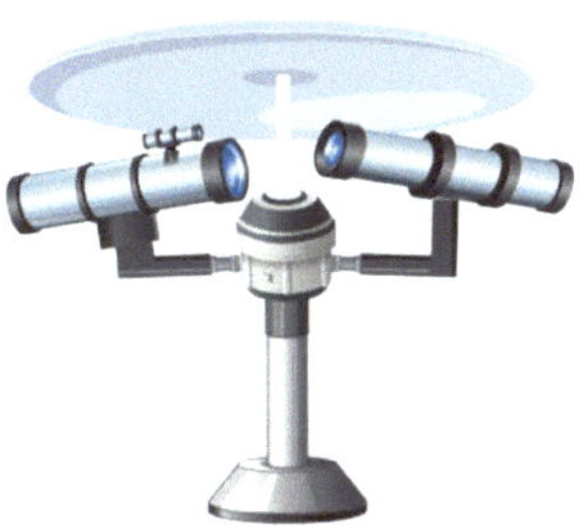

The Francium atom is useful for precision testing of relativistic atomic physics—the study of the structure of an atom. It provides valuable insights into the behavior of heavy and radioactive elements.

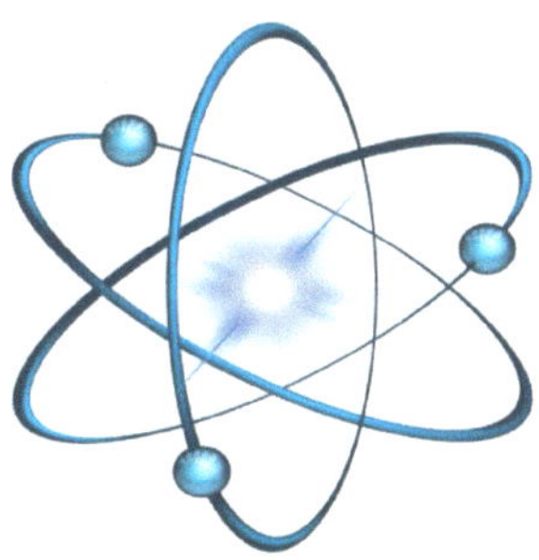

Scientists use Francium in spectroscopic experiments to study the tiny parts of things. They look at how things are made up and how they work on a small scale and even in space.

Some people on the internet are very excited about something called Francium Bombs. They think that the Francium element would make a huge explosion when it touches water. But the truth is scientists probably won't ever be able to make enough Francium that lasts long enough to actually build a bomb.

Did You Know?

- Only 4 women have discovered elements of the periodic table. Francium was discovered by one of them - Marguerite Perey - a nuclear chemist in France. The initial element symbol was Fa (1946) before it changed to Fr (1949).
- Ms. Perey wanted to name the element catium as she believed it to be the most electropositive cation of the elements. But her bosses didn't like it because they thought people would think about cute little cats. So, they decided to name it after the country, France, instead.
- Francium is an element that we can't see because it is made from the break down of so many other elements. It's almost impossible to get a sample of this element because it gets super hot and disappears too quickly for us to see. Scientists think it would be a very shiny liquid at a warm room temperature. Francium is the last naturally occurring element to be discovered.

Gallant Presents Gallium

Symbol: Ga Atomic Number: 31 Atomic Mass: 69.72

Gallium Facts

- Discovered in 1875 in Paris
- Soft, solid metal
- Diamagnetic in solid state. Magnetic as liquid.
- Excellent conductor of heat and electricity
- Is both ductile and malleable
- Post-Transition Metal

Gallium is obtained mainly from mining and mineral processing of bauxite ore for aluminum. The largest producers of Gallium are China, Australia, Russia, France, and Germany.

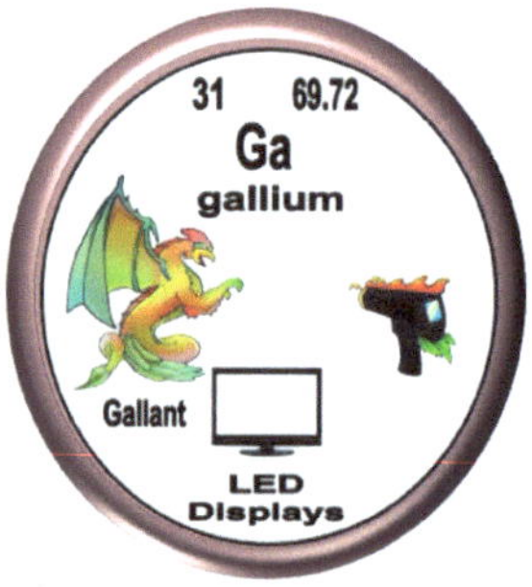

Gallant's Magical Abilities

- When Gallant aims an ion beam of Gallium from the tip of his tail at anything made from aluminum it causes something called "gallium-induced structural failure" to occur. The aluminum will melt to gray goo rendering it useless.

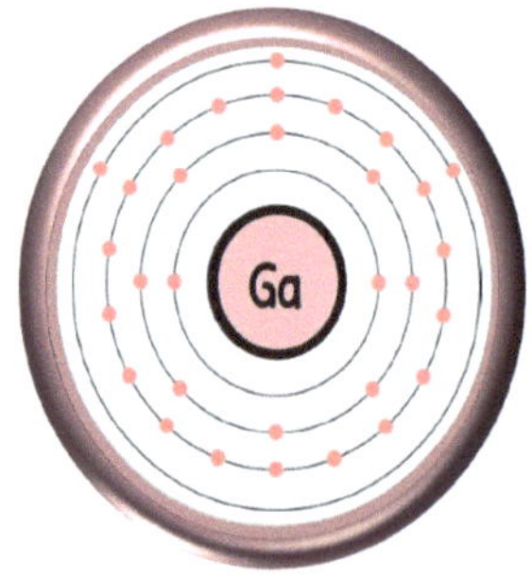

Atomic Structure

Uses For Gallium

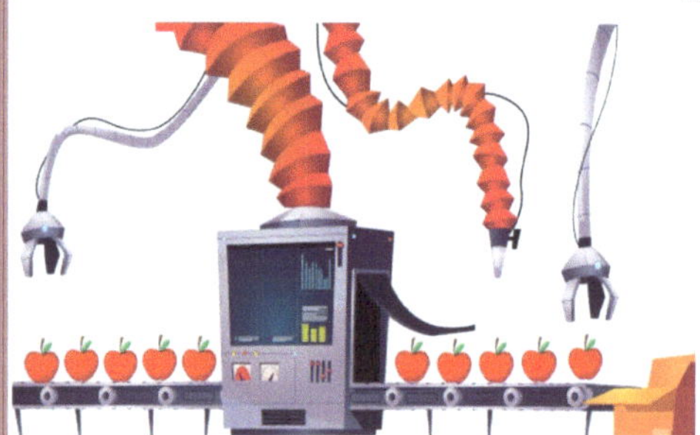

Gallium is one of 4 metals that melt at room temperature. It is the least reactive and least toxic, making it the most safe and environmentally-friendly choice for high temperature thermometers.

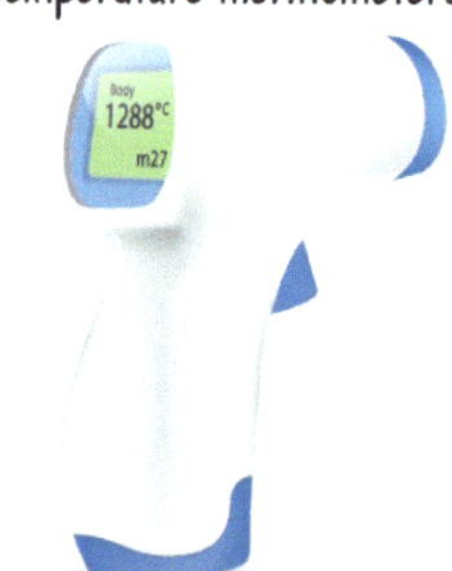

Short-wave infrared (SWIR) cameras create thermal images, and can image through glass to inspect anything from solar cells to produce.

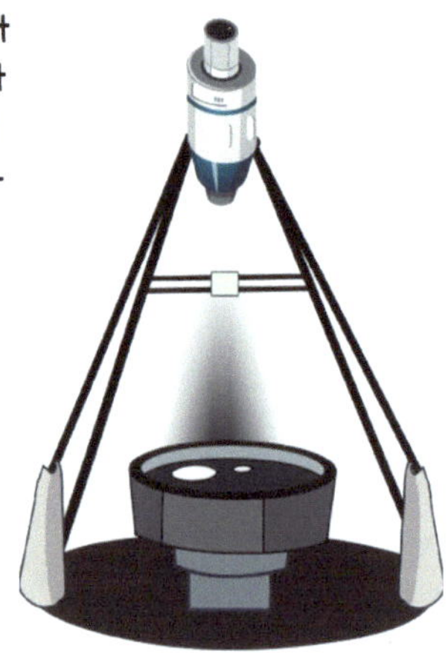

A Gallium-indium alloy is used for liquid mirror reflecting telescopes. Astronomers plan to some day set up a spinning vat of reflective liquid (like Gallium) in a moon crater to investigate the very first stars.

Did You Know?

- Lecoq de Boisbaudran named Gallium after France ('Gallia' in Latin) and also after himself. Lecoq, which means 'the rooster' translates to 'Gallus' in Latin. The pictorial symbol for Gallium is a silvery metallic rooster shown on a background of an antique map of France.
- Like Mercury, a blob of liquid Gallium in water can simulate a "beating heart". It will easily melt from the warmth of your hand but it is very hard to boil. You would need to heat this metal to 3,999°F (2,204°C) to make it boil.
- Carrying Gallium on aircraft is heavily restricted due to the fact that Gallium-Induced Structural Failure can occur in anything made from Aluminum.

Hildy Presents Hydrogen

Symbol: H Atomic Number: 1 Atomic Mass: 1.008

Hydrogen Facts

- First artificially produced in the early 16th century
- The lightest gas
- Poor conductor of heat and electricity
- Very weakly magnetic— Diamagnetic
- H2 (molecular Hydrogen) is highly reactive
- Non-Metal Gas

Green Hydrogen is produced by using electricity to split water molecules (H2O) into hydrogen (H2) and oxygen (O2). Blue hydrogen is produced from hydrocarbon sources, such as natural gas, and combined with carbon capture technology.

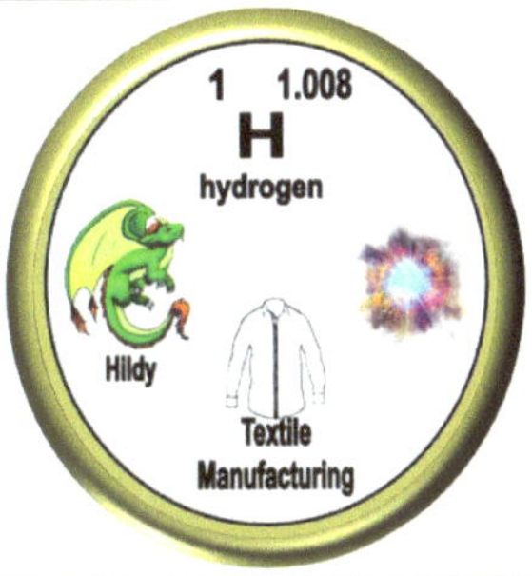

Hildy's Magical Abilities

- A rainbow blast of energy from Hildy's Hydrogen tipped tail provides super fuel to power vehicles from race cars to space craft.

Atomic Structure

Uses For Hydrogen

Hydrogen is used to fuel vehicles in some countries.

Hydrogen is used in the hydrocracking process which takes heavier refinery products like crude oil and "cracks" the large molecules into smaller ones to make things like gasoline, diesel and thousands of other chemicals.

Next to oil refineries, ammonia is currently the largest application of hydrogen. Ammonia (NH3) is used to produce ammonium nitrate, a fertilizer, and is part of many household cleaning products. Don't forget to wear your gloves when using it!

Hydrogen is used to turn unsaturated fats into saturated oils and fats through hydrogenation. Hydrogen is used to make hydrogenated vegetable oils in foods such as margarine, peanut butter, and it is even in some skin care products.

Did You Know?

- Henry Cavendish, during the years 1766-1781, was the first person to find out that hydrogen gas turns into water when it is burned. Because of this, it got its name from the Greek word "hydrogen" which means "water-former".
- Hydrogen is color coded. There's green, blue, grey, black and brown, pink, turquoise, yellow and white. Green is the cleanest and black is the most environmentally damaging.
- Hydrogen is not like the noble gases because it only has one electron in its shell, which makes it unstable. Unlike noble gases that have full outer shells of valence electrons, hydrogen is different. When hydrogen gas gets activated by heat, light, or other things, it reacts very fast and releases heat. It can also combine with oxygen and other reactive gases to form harmless water vapor.

Iker Presents Indium

Symbol: In Atomic Number: 49 Atomic Mass: 114.8

Indium Facts

- Discovered in 1863 in Germany
- Softest non-alkali metal
- Good conductor of heat and electricity
- Diamagnetic (repelled by a magnetic field)
- Highly ductile and malleable
- Post Transition Metal

Some indium is found in pure form, in the mineral, indite. Most indium is obtained as a by-product of zinc processing and is recovered from residues left from electrolytic refining of zinc.

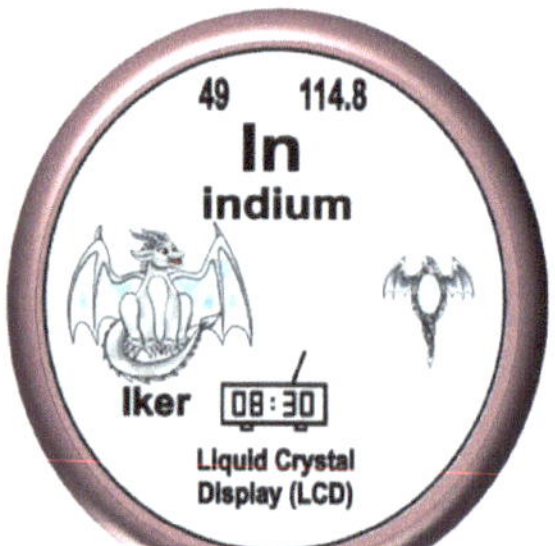

Iker's Magical Abilities

- A touch of Iker's Indium tipped tail causes any surface to become a reflective mirror.

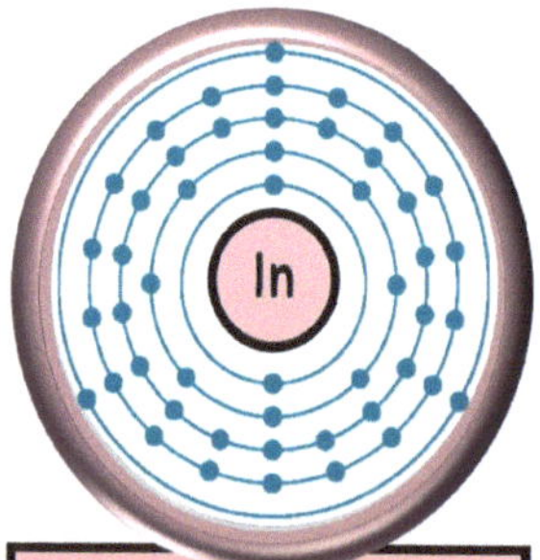

Atomic Structure

Uses For Indium

Thin coatings of ITO over the windshields of aircraft or cars, allows the glass to de-ice or de-mist, and it can reduce air conditioning requirements.

Alloys that contain gallium, indium, and tin are called Galinstan, and are liquid at room temperature. It is an alternative to thermometers that use mercury which is much more toxic.

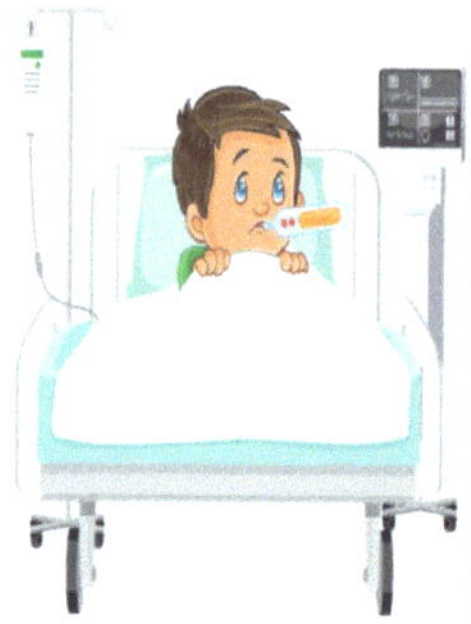

Indium tin oxide (ITO) is the best material to fill the growing need for LCD's (liquid crystal displays) in touch screens, flat screen TVs and solar panels.

Indium's low-melting alloys are also used in sprinkler heads, fire-door links, and fusible plugs.

Indium is used in street lights to reflect infrared light back to the light bulb,

Did You Know?

- Indium has the unusual property when molten of clinging to (wetting) clean glass and other surfaces. This is helpful for sealing things like glass, metals, and ceramics together really tightly.
- Indium is an amphoteric element that can act as both an acid and a base. It can dissolve in acid to make indium salts, and it can dissolve in strong alkalies to make indium compounds called indates.
- In May 2018, the U.S. government made a list of 35 important minerals, and indium is included. This list was created to make sure we can always get these minerals when we need them. To learn more, see Executive Order 13817, "A Federal Strategy to Ensure Secure and Reliable Supplies of Critical Minerals" (82 FR 60835).

Jalan Presents Jazzy Neon

Symbol: Ne Atomic Number: 10 Atomic Mass: 20.18

Neon Facts

- Discovered 1898 in London
- Monoatomic "single atom" gas
- Insulator—Not a good conductor of heat and electricity
- Colorless - Glows Red
- Unreactive (and stable)
- Noble Gas

Neon is produced from liquid air. Liquid air is air that has been cooled to very low temperatures in cryogenic air-separation plants, so that it has condensed into a pale blue mobile liquid.

Jalan's Magical Abilities

- With a swish of her Neon tipped tail, Jalan can paint brightly colored light patterns on surfaces that last for days.

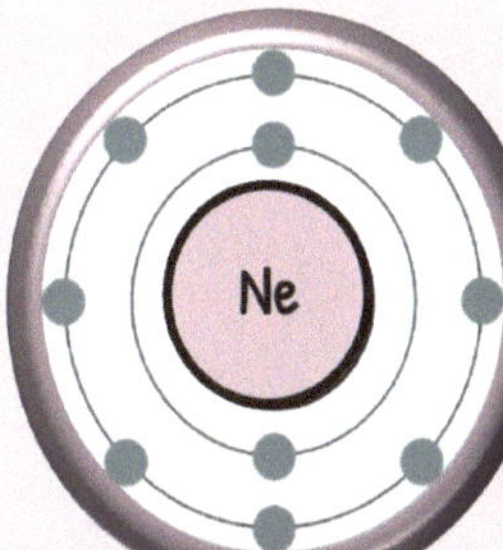

Atomic Structure

Uses For Neon

Neon is used in High Voltage Indicators to indicate the presence of voltage up to 15kV to ensure the safety of maintenance or installation personnel allowing the apparatus to be safely grounded .

Neon is used in lightning arresters—devices installed to shield power lines, homes, and structures from dangerous power surges.

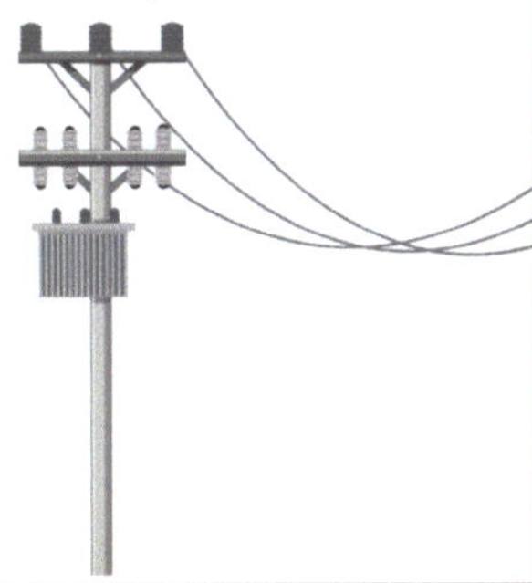

Neon is used in cryocoolers - machines that keep sensors extremely cold so that satellites and space telescopes (like the James Web above) can take pictures on long-term missions.

Pure Neon signs are reddish-orange color. Other colors are produced when a number of other gases like Mercury (blue), Helium (pink or yellow), Xenon (lavender), Argon (light blue or purple) or Krypton (whitish yellow) are used.

Did You Know?

- The largest use of neon is in making 'neon signs' for advertising. Pure neon gas glows red but Georges Claude, the discoverer, found that when neon gas was combined with other elements and fed an electrical charge different brightly colored lights were displayed within the sealed tube.
- Most neon signs have either neon or mercury vapor in them, depending on whether the manufacturer wants to create colors with a cool or warm base color. When a manufacturer wants cool colors, like green or purple, they will use mercury rather than neon.
- Neon is 40 times more effective as a refrigerant than liquid helium and three times better than liquid hydrogen. Because of its high refrigeration capacity, liquid neon is used in cryonics to freeze corpses for preservation or for potential revival in the future.

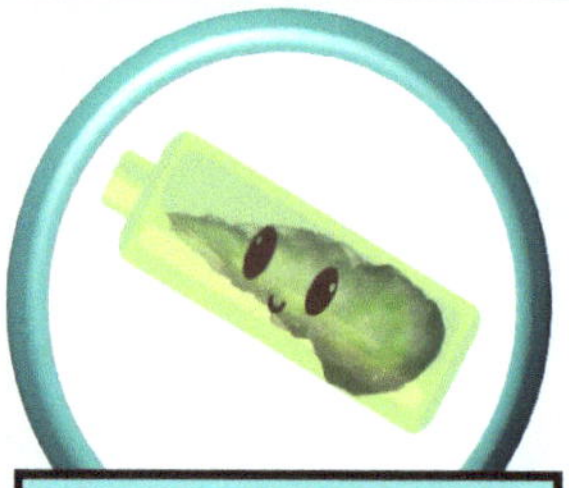

Krystix Presents Keen Chlorine

Symbol: Cl Atomic Number: 17 Atomic Mass: 35.45

Chlorine Facts

- First recognized as a gaseous substance in 1630 in Brussels
- 2nd lightest Halogen gas
- Not a conductor of electricity – Insulator
- Diamagnetic
- Highly Reactive
- Halogen

The main source that Chlorine comes from is a salt called Halite (Sodium Chloride - NaCl). It is found in very dry places with lots of salt flats - like deserts.

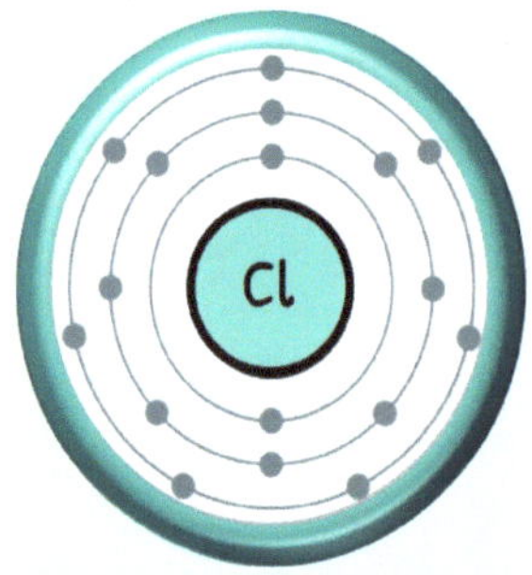

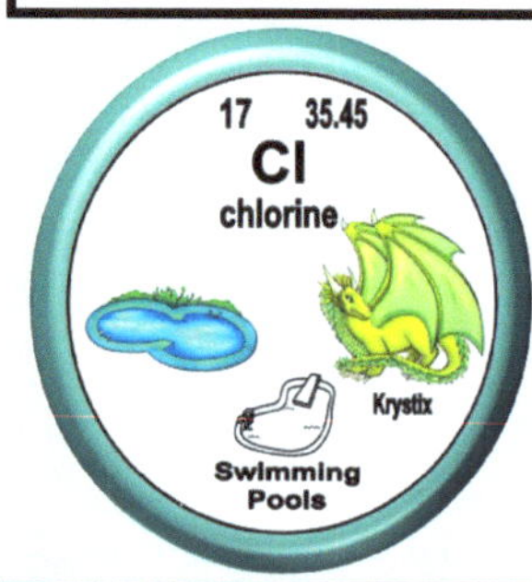

Krystix's Magical Abilities

- The tiniest tap of Krystix's Chlorine tipped tail purifies any water it comes in contact with.

Atomic Structure

Uses For Chlorine

Chlorine Bleach is the most used disinfectant for cleaning and whitening agent for fabrics.

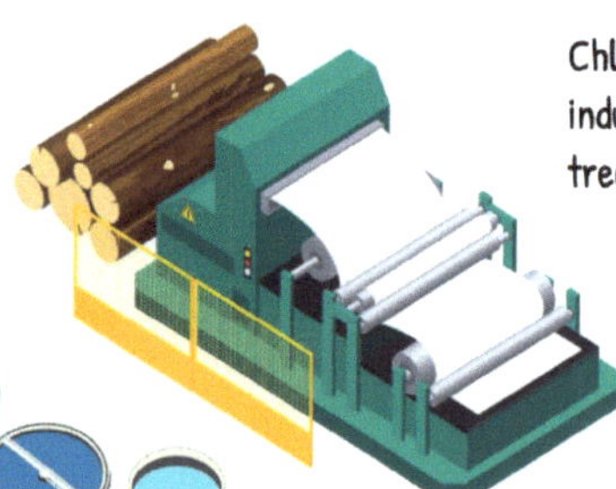

Chlorine is used in the pulp and paper industry to break down the woody cell walls of tree pulp so that it can be made into paper.

Chlorine is part of the sanitation process for sewage and industrial waste.

Chlorine is used to purify water because it kills pathogens such as bacteria and viruses by breaking the chemical bonds in their molecules.

Did You Know?

- The true nature of Chlorine gas as an element was recognized in 1810 by English chemist Humphry Davy, who later named it chlorine. The Greek word for "greenish-yellow" is khloros.
- More than 2,000 naturally-occurring chlorine-based compounds have been identified in all kinds of organisms and places, including the human body.
- Chlorination is one of many methods that can be used to disinfect water. This method was first used over a century ago, and is still used today to combat waterborne diseases. The use of Chlorine has proven effective against bacteria and viruses and makes our water safer.

Lillian Presents Lithium

Symbol: Li Atomic Number: 3 Atomic Mass: 6.94

Lithium Facts

- First detected in 1817 in Stockholm
- More brittle than other alkali metals
- Good conductor of heat and electricity
- Not inherently magnetic
- Ductile and malleable
- Alkali Metal

Lithium is extracted from Spodumene and other ores like petalite and lepidolite. It's also found in great quantities in salt water. Australia and Chile are the world's largest producers of lithium.

Lillian's Magical Abilities

- A soft caress from Lillian's Lithium tipped tail quiets and calms chaotic feelings.

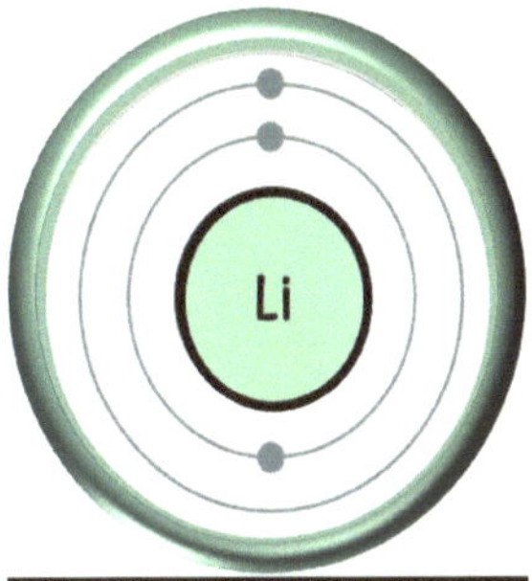

Atomic Structure

Uses For Lithium

Small rechargeable Lithium batteries are used in pacemakers, digital cameras, smart phones, laptops, watches and more. Large Lithium batteries are used in scooters, golf carts, trolleys, boats, cars, trucks, recreational vehicles and more.

Lithium is a medication that effectively reduces manic symptoms and is a treatment for bipolar episodes.

Lithium is used to insulate windows from the sun.

Lithium is an essential ingredient for some specialty glasses and glass-ceramics like glass cook-top stoves. It is responsible for the glass-ceramic's low or near zero thermal expansion, enabling their use in high-temperature ranges without voltage breakage.

Did You Know?

- Certain lithium compounds, also known as lithium salts, are used as psychiatric medication, primarily for bipolar disorder and for major depressive disorder. Lithium citrate was part of the 7 Up soft drink recipe until 1948 when the U.S. government forced American beverage makers to remove all lithium from their recipes.
- Lithium is one of three elements produced by the big bang, along with hydrogen and helium.
- Pure lithium metal is extremely corrosive and requires special handling. Because it reacts with air and water, the metal is stored under oil or enclosed in an inert atmosphere. When lithium catches fire, the reaction with oxygen makes it difficult to extinguish the flames.

Maximo Presents Molybdenum

Symbol: Mo Atomic Number: 42 Atomic Mass: 95.95

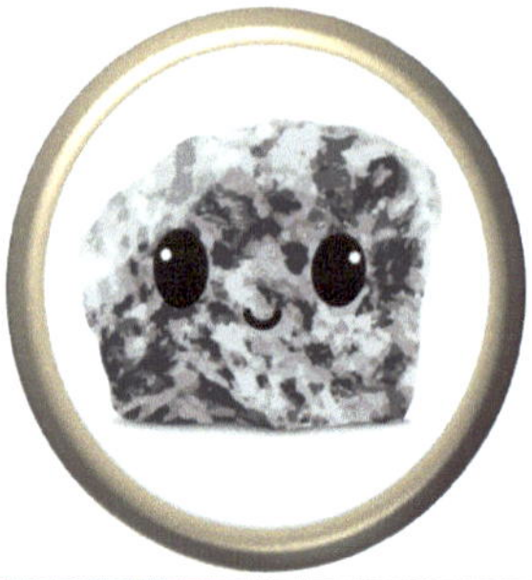

Molybdenum Facts

- Discovered in 1781 in Sweden
- Shiny Silver White Solid
- Good conductor of heat and electricity
- Paramagnetic
- Somewhat ductile and malleable
- Transition Metal

The main Molybdenum ore is Molybdenite (molybdenum disulfide). It is a by-product of copper or tungsten mining and is mined primarily in the United States, China, Chile and Peru.

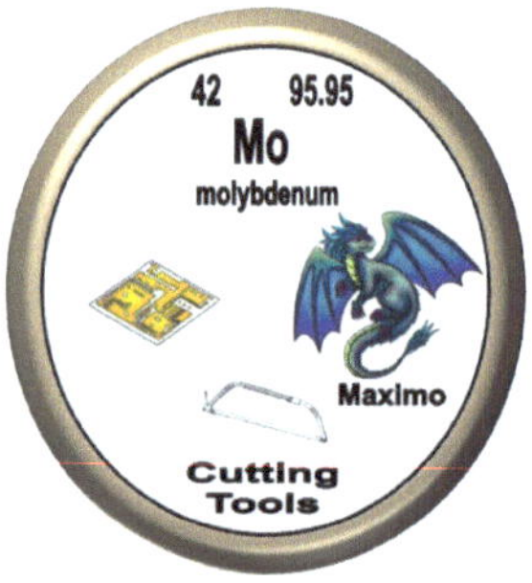

Maximo's Magical Abilities

- A swish of Maximo's Molybdenum tail permanently bonds metal to glass.

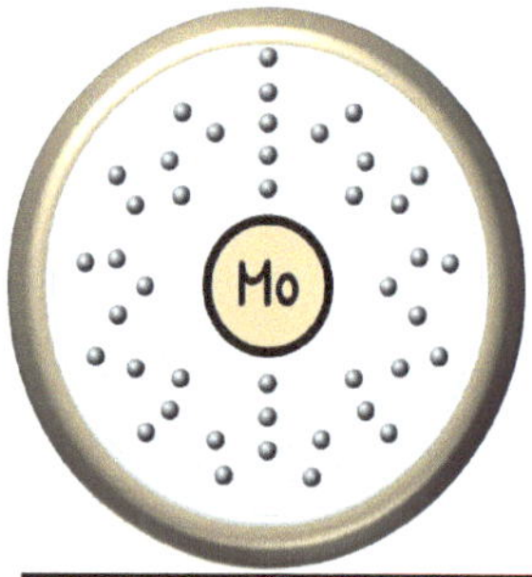

Atomic Structure

Uses For Molybdenum

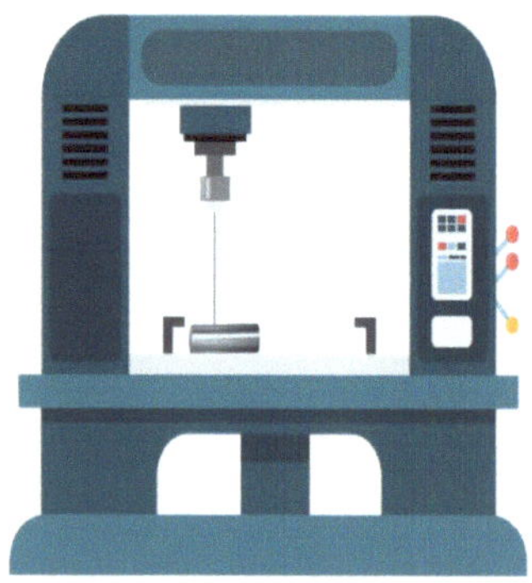

Wire electrical discharge machines (wire EDM) use molybdenum wires to cut through titanium and other hard metals.

Adding Molybdenum Lanthanum to steel makes it stronger, tougher, last longer against rust, and it doesn't wear out easily.

Seeds using ammonium molybdenum additives can greatly increase the yield of flowers.

Molybdenum is an essential trace mineral that occurs naturally in foods.

Did You Know?

- Although Molybdenum was found in the late Eighteenth Century, it had been used before then. In the Fourteenth century, the Japanese used Molybdenum steel knives and swords. In the sixteenth century, Molybdenite was similar to Lead, Galena, and Graphite in its appearance and properties, and was even used as graphite. There is no actual graphite in Molybdenum, though.
- Molybdenum got its name from Molybdenite, which got its name from the Greek "molybdos," which means "lead." There is no Lead in Molybdenum.
- Molybdenum is one of the five major refractory metals (metals with very high resistance to heat and wear). The others are tungsten, tantalum, rhenium and niobium.

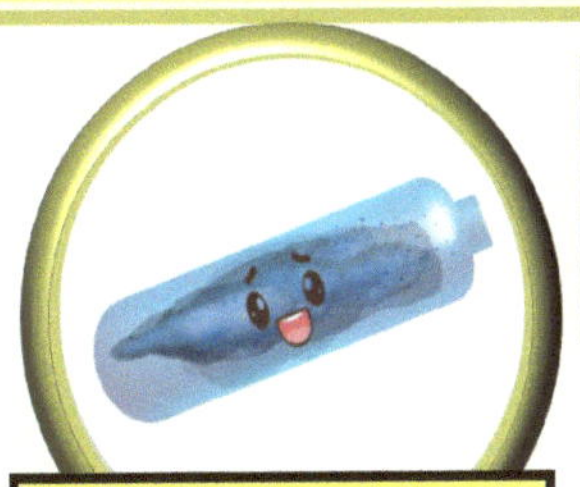

Nitra Presents Nitrogen

Symbol: N Atomic Number: 7 Atomic Mass: 14.01

Nitrogen Facts

- Officially discovered in 1772 by Scottish physician D. Rutherford
- Colorless, odorless, tasteless, and inert gas
- Poor conductor of heat and electricity
- Diamagnetic
- Neither ductile or malleable
- Non-Metal—Gas

The largest reservoir of Nitrogen is found in the atmosphere. Nitrogen gas makes up 78% of the air we breathe. The three standard methods used to extract nitrogen from air are cryogenic distillation, pressure swing adsorption and membrane nitrogen generation.

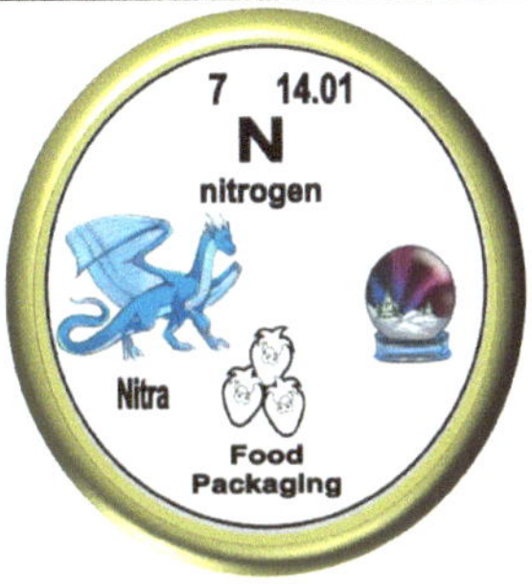

Nitra's Magical Abilities

- A gentle tap of Nitra's Nitrogen tipped tail causes happy feelings to bubble to the surface of anyone she touches.

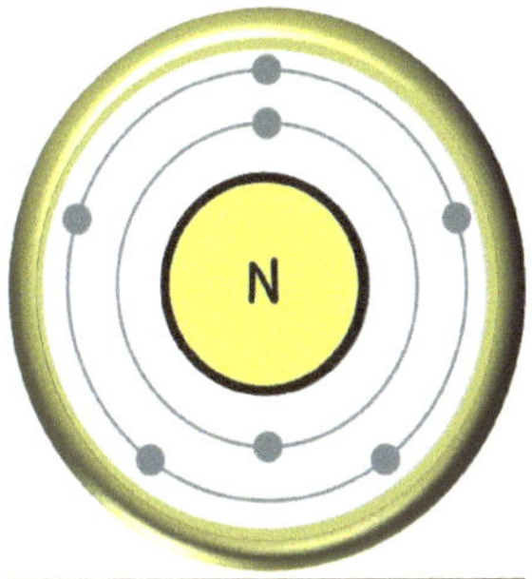

Atomic Structure

Uses For Nitrogen

Nitrogen is used to create nylon fabrics that are used for many things including parachutes.

Nitrogen gas is used to preserve the freshness of packaged or bulk foods like chips, popcorn, nuts, and prepackaged lunch meat.

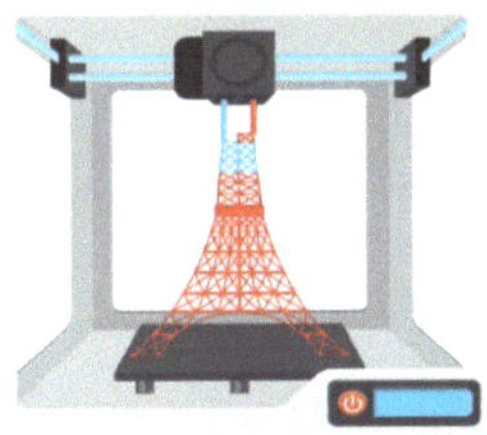

Nitrogen is commonly used to purge oxygen from 3D printed parts.

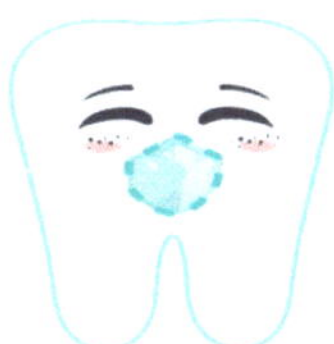

Nitrous Oxide, an oxide of Nitrogen (N_2O) is sometimes used by dentists to help patients feel less anxious about their dental procedure.

Did You Know?

- In the 1800s, British upper class had laughing gas parties where they inhaled the compound Nitrous Oxide, also known as laughing gas. This trend also reached the United States, where its medical benefits were quickly recognized. Even today, dentists and other medical professionals use it to reduce anxiety.
- Manure contains a lot of Nitrogen. A long time ago, farmers realized that when they kept their animals and plants together, it made the plants grow super strong. There's proof people started using manure for growing legume crops more than 6,500 years ago.
- Nitrogen gas has a significant role in the formation of an aurora. The stunning hues of purple, blue, or pink that we see in the aurora borealis are a result of Nitrogen's presence.

Ozzy Presents Oxygen

Symbol: O Atomic Number: 8 Atomic Mass: 16.00

Oxygen Facts

- Discovered in 1774 in England
- Colorless, odorless, tasteless gas
- Poor conductor of heat and electricity
- Paramagnetic
- Neither ductile or malleable
- Non-metal Gas

Roughly half of the Oxygen production on Earth comes from the plankton in the ocean. Commercial-grade Oxygen is primarily produced through gas separation of air via cryogenic distillation in an air separation unit (ASU).

Ozzy's Magical Abilities

- A tap of Ozzy's Oxygen tipped tail instantly expands the size of a living object 10 times.

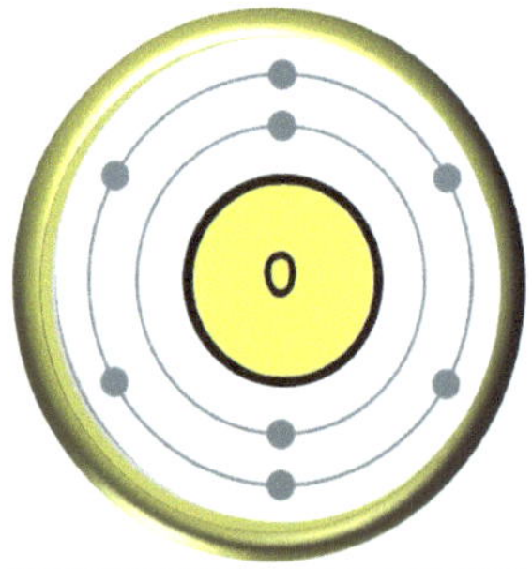

Atomic Structure

Uses For Oxygen

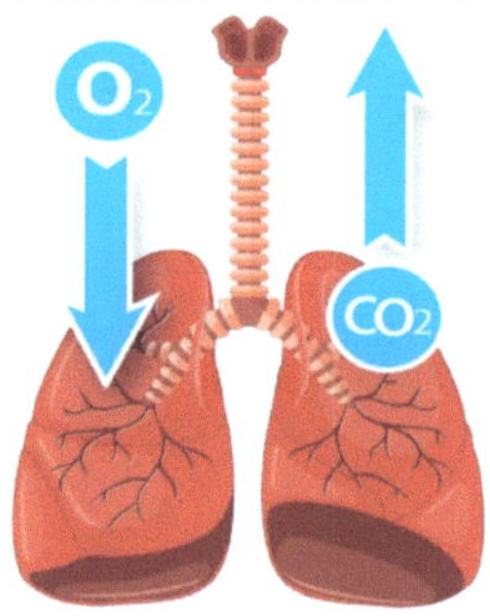

We all require oxygen to live. The respiratory system's main job is to transport oxygen through the body and remove extra carbon dioxide.

Ozone (O3) is an allotropic form of oxygen that is added in small amounts to breathing air for its invigorating effect and "fresh air" scent.

Aquaculture, the cultivation of fish in ponds, uses oxygenated water to ensure sufficient oxygen is always present.

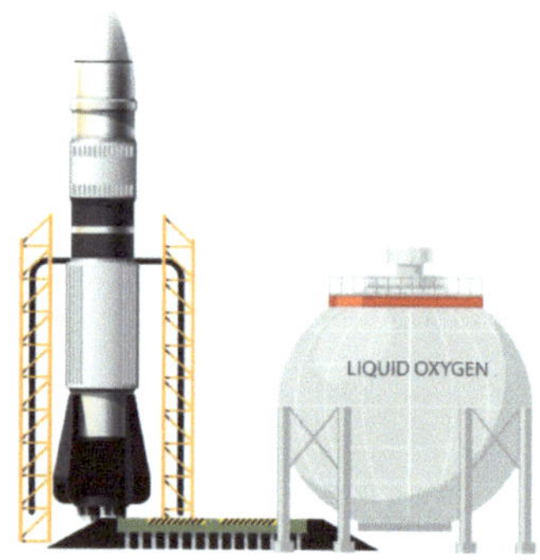

Liquid Oxygen (LOX) is stored in spheres with cryogenic refrigeration systems. One of LOX's main uses is rocket fuel.

Did You Know?

- 2,500 years ago, the Greeks identified air – along with earth, fire and water – as a fundamental component of the world. Oxygen is vital for life, making up 20.94% of the air we breathe. It is colorless, odorless, and tasteless but essential for supporting life on Earth.
- Glowing oxygen is generated on Earth during polar auroras through the interaction of energetic electrons from space and the upper atmosphere. This emission of light, fueled by oxygen, is what gives polar auroras their distinct and stunning green color.
- When you heat up a drop of liquid oxygen in the air, it gets much bigger. It turns into 860 times more oxygen gas than before! So, it expands a lot!

Paedra Presents Platinum

Symbol: Pt Atomic Number: 78 Atomic Mass: 195.1

Platinum Facts

- First written account in 1557
- Somewhat brittle, solid
- Good conductor of heat and electricity
- Paramagnetic
- Highly ductile and malleable
- Noble Transition Metal

Most commercially produced Platinum comes from South Africa, from the mineral cooperite (platinum sulfide). Some platinum is prepared as a by-product of copper and nickel refining.

Paedra's Magical Abilities

- When Paedra swishes her shiny Platinum tipped tail in the air, it turns dangerous gases into safer things, like water vapor.

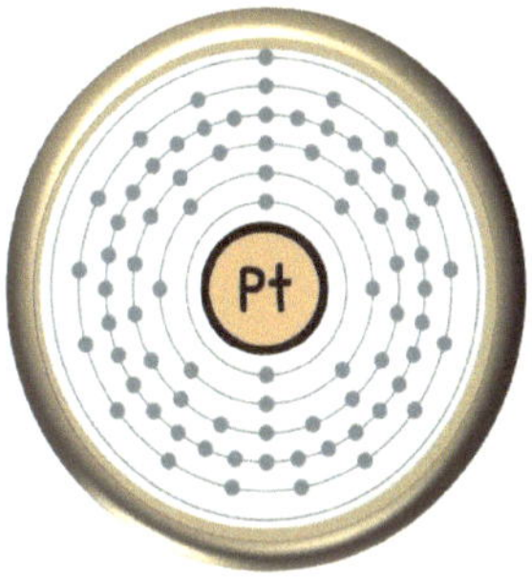

Atomic Structure

Uses For Platinum

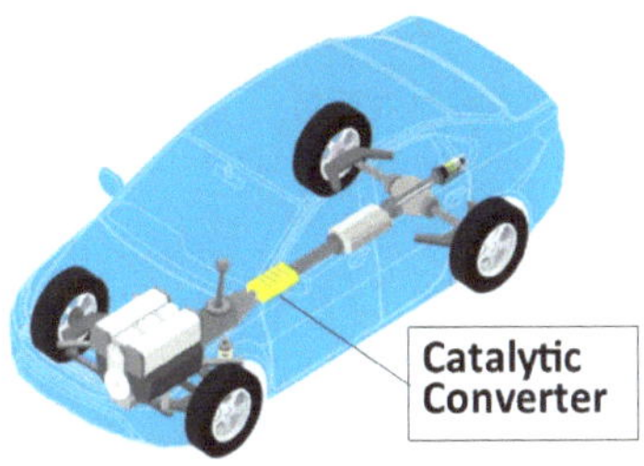

Platinum is used in catalytic converters, which change harmful substances in a car's exhaust gasses, such as carbon monoxide, nitric oxide, nitrogen dioxide and hydrocarbons, into less harmful substances like carbon dioxide and water vapor by means of chemical reactions.

Platinum-cured silicone mixtures are used in cosmetics to make skin look better. They can help prevent dryness and make wrinkles and lines on the face less noticeable. They also help make skin look younger.

Over 30% of platinum is used for jewelry. The natural bright white color, satisfying weight, hardness, and resistance to tarnishing makes it an ideal metal for jewelry.

Platinum catalysts are used to make petrochemical feed stocks, the basic raw materials for the manufacture of plastics, synthetic rubber and polyester fibers used to produce clothes and blankets.

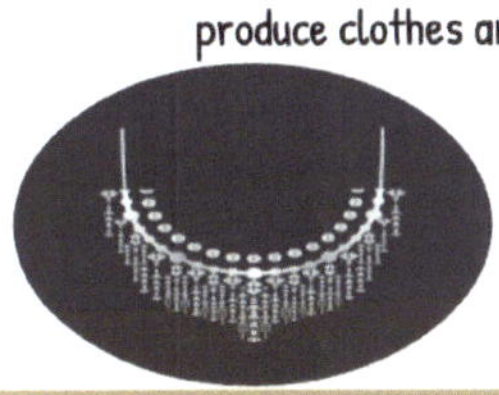

Did You Know?

- Catalytic converters use elements like Platinum (Pt), Palladium (Pd), and Rhodium (Rh) as catalysts to reduce nitrogen oxides (NOx) in car exhaust .
- In 1814, the United States struck a half dollar coin in platinum. As a form of currency, these coins proved to be impractical: platinum resembles many less expensive metals, and, unlike the more malleable and ductile silver and gold, it is very difficult to work. Today platinum coins are only used as investments or treated as collector coins.
- During World War II, platinum was not allowed to be used in jewelry production because it was such a vital component in strategic defense engineering. A six-inch cube of platinum is so dense it weighs the same as the average adult (about 165 pounds).

Quincy Presents Quicklime

The chemical formula of the compound is CaO.

Quicklime Facts

- Egyptian pyramids contain Lime plastering
- White, caustic, crystalline solid
- Not a conductor of heat and electricity
- Not Magnetic
- Neither ductile or malleable
- Alkaline Compound

Quick Lime comes from the thermal decomposition of limestone. It is also known as "burnt lime". The chemical formula of the compound is CaO.

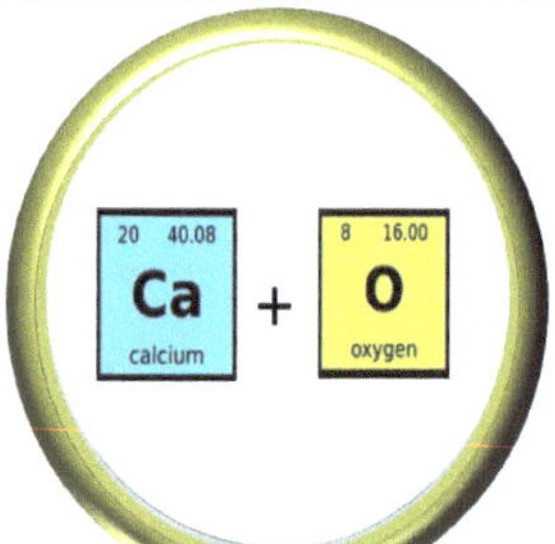

Quincy's Magical Abilities

- Quincy can use his Quicklime tipped tail to direct a beam of intense light at some one or thing, effectively placing them in the "limelight".

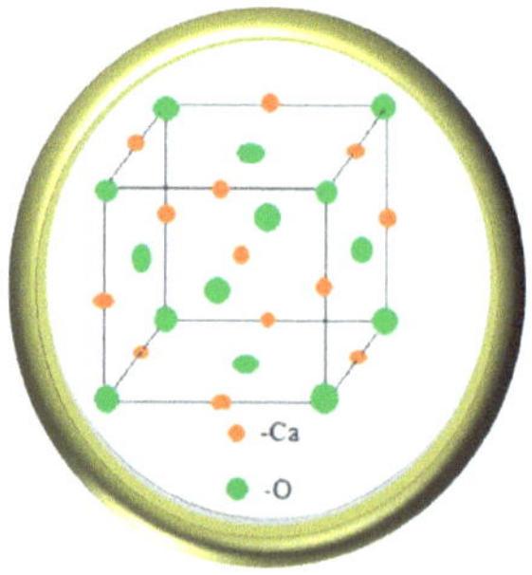

Quicklime Formula

Uses For Quick Lime

Cement, made with Quick Lime, is probably the most used man-made material. It is used for building everything from architectural structures to highways and dams.

Quicklime was used to create limelight, a popular way to illuminate theater stages in the Nineteenth Century.

440 lbs of lime is used per ton of sugar beets to force out any impurities in raw sugar juice. It acts as a clarifying agent for sugarcane juice in the manufacture of white sugar.

Quicklime is the most effective and cost efficient alkaline product used in the making of drinking water, as well as in the treatment of waste water and sludge. In fact, most of the quicklime that is produced is used to improve the quality of drinking and industrial water.

Did You Know?

- Calcium oxide (CaO) is usually known as burnt lime or quicklime.
- The famous Pantheon in Rome boasts the world's largest unreinforced concrete dome—an architectural marvel that has endured for millennia, thanks to the incredible durability of ancient Roman concrete.
- Limelight (also known as Drummond light or calcium light) is a non-electric type of stage lighting once used in theatres and music halls, and was the first solid-state lighting device. Although it has long since been replaced by electric lighting, the term has nonetheless survived, as someone in the public eye is still said to be "in the limelight".

Ruth Presents Ruthenium

Symbol: Ru Atomic Number: 44 Atomic Mass: 101.1

Ruthenium Facts

- Discovered in Russia in 1844
- Very brittle, hard, white metal
- Good conductor of heat and electricity
- Magnetic at room temp
- Malleable, ductile and sonorous
- Transition Metal

Ruthenium comes from pyroxenite deposits in South Africa and other places. It is also found in ores with the other platinum group metals in the Ural Mountains and in North and South America.

Ruth's Magical Abilities

- Ruth can wave her Ruthenium tipped tail over a surface to expose fingerprints and other oily markings.

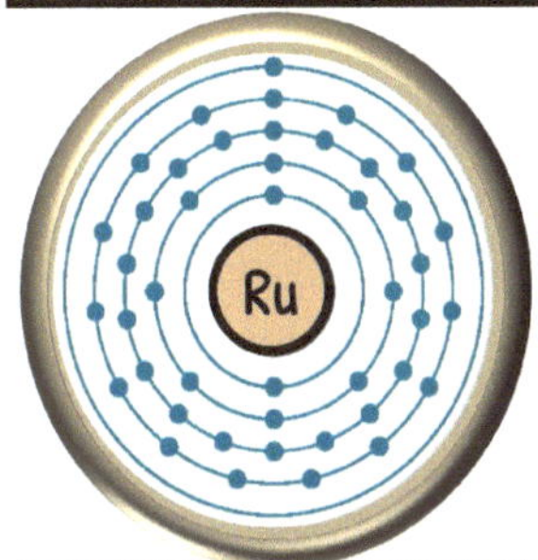

Atomic Structure

Uses For Ruthenium

Ruthenium Tetroxide- Ruthenium Tetroxide (RuO4) is used to find fingerprints on different things like plastic, doors, and skin. It works on both porous and non-porous surfaces.

Ruthenium-based dyes are used to manufacture Dye-sensitized Solar Cells (DSCs), 3rd generation solar cells that provide high efficiency light-to-electricity conversion with low production costs.

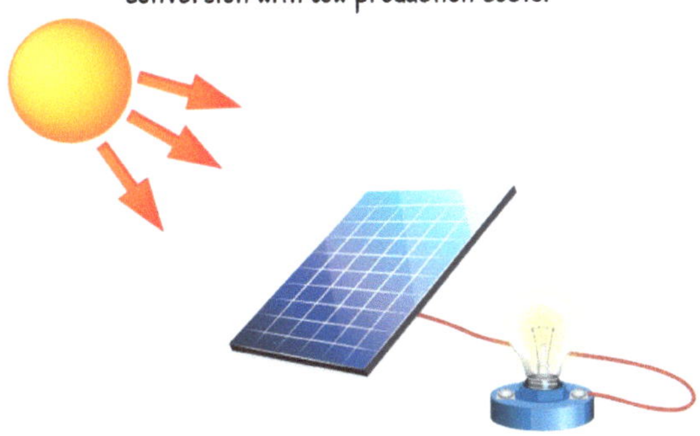

Computer hard disk drive memory using Ruthenium as an underlayer to its Platinum-containing storage layer allows more information to be stored in a smaller space. This means we can save a lot more stuff on our computers!

Did You Know?

- Ruthenium (Ru) has magnetic properties at room temperature. Ru is only the fourth element to have this property; the others are iron, cobalt, and nickel.
- Ruthenium is a sonorous metal, meaning it makes a sound when it gets hit. It's like the metals used to make bells, violins, and guitars. Some examples of sonorous metals are: Copper, Tin, Steel, Nickel, Silver and Iron.
- One place where you might have come across Ruthenium in everyday life is jewelry: the metal's hardness, high corrosion resistance and unusual, not-quite-metallic grey-black finish make it popular choice. Pure ruthenium is expensive though, so it's almost always plated onto a cheaper base metal.

Sorn Presents Sodium

Symbol: Na Atomic Number: 11 Atomic Mass: 22.99

Sodium Facts

- Discovered in 1807– London
- Soft, silvery-white solid, highly reactive
- Excellent conductor of heat and electricity
- Paramagnetic
- Ductile and malleable
- Alkali Metal

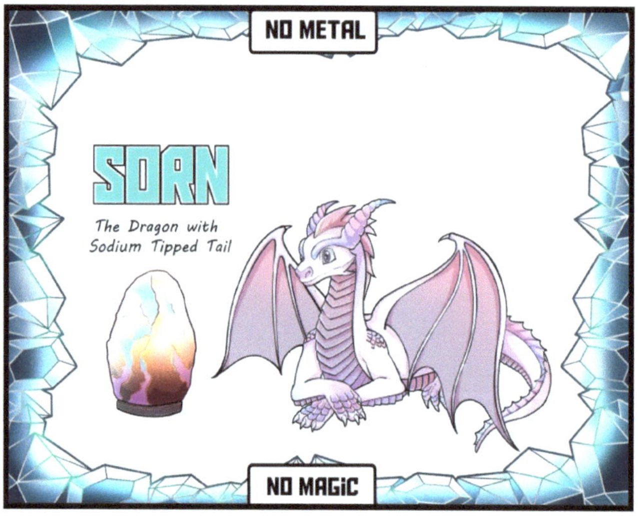

Traditionally, salt was harvested as sodium chloride from either solar evaporation ponds, or rock or clay deposits. It is mined in nearly every country in the world.

Sorn's Magical Abilities

- When Sorn rests his Sodium tipped tail on an animate or inanimate surface it drastically lowers its temperatures.

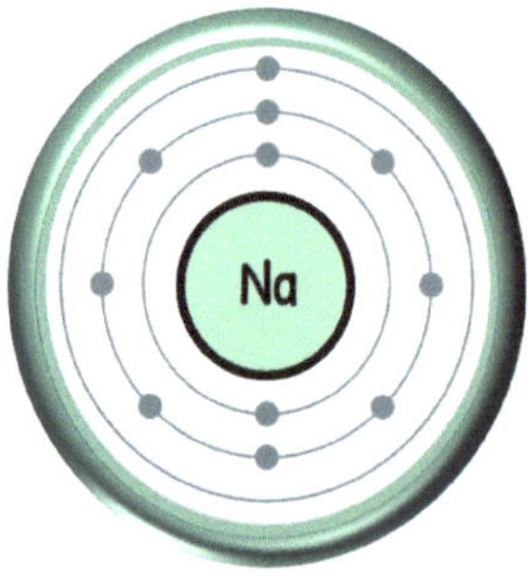

Atomic Structure

Uses For Sodium

Adding salt to ice brings its temperature down below freezing. It lowers the temperature from the normal freezing point of water (0 °C or 32 °F) to as low as -21 °C or -5 °F. That's cold enough to make ice cream.

Some Chinese companies are using Sodium to make batteries as its chemical properties are very similar to that of Lithium.

Sodium is used in the synthesis of perfumes. Perfumers use salt to add contrast to heavier, sweeter scents.

Sodium Chloride is used to de-ice roads and sidewalks. Road salt works by lowering the freezing point of water, causing ice to melt even when the temperature is below water's normal freezing point of 32 degrees.

Did You Know?

- The chemical symbol for sodium is Na, which is abbreviated from the Latin word, Natrium. Sodium actually comes from the Arabic word, suda, meaning headache. Sodium carbonate was used in Arabic culture as a headache remedy.
- Sodium is an essential element in animal nutrition. In humans, sodium is important for maintaining fluid balance in the cells and throughout the body, while the electric potential maintained by sodium ions is critical for nerve function. If too much sodium is ingested, your body will try to excrete it as soon as possible via urine, sweat and tears.
- Sodium is one of only three metals light enough to float on water; the other two being lithium and potassium.

Tilly Presents Titanium

Symbol: Ti Atomic Number: 22 Atomic Mass: 47.87

Titanium Facts

- Discovered in 1791 in the Manaccan Valley
- Hard but brittle, shiny solid
- Poor conductor of heat and electricity
- Diamagnetic
- Ductile or malleable
- Transition Metal

Titanium ore is mined from intrusive crystalline rocks, weathered rock and unconsolidated sediments known as shoreline placer deposits. Placers are alluvial deposits formed by rivers as they reach the sea.

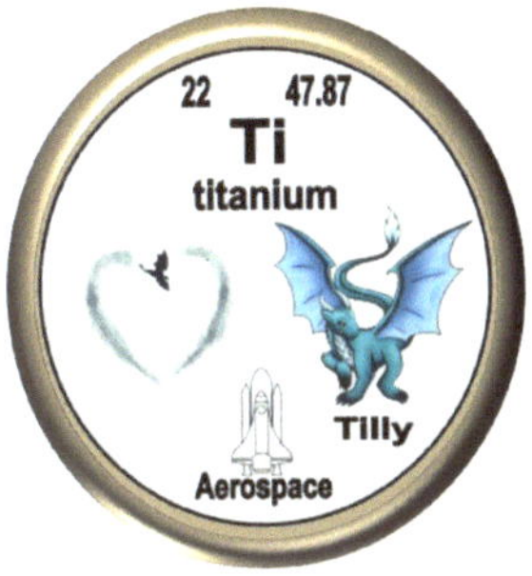

Tilly's Magical Abilities

- Tilly can blast a thick cloud stream into the air from her Titanium tipped tail to write words across the sky.

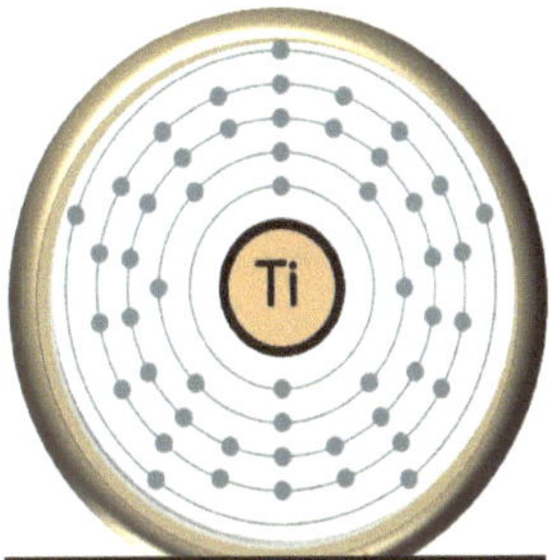

Atomic Structure

Uses For Titanium

Because Titanium is bulletproof and bullet resistant, it is a popular material used by the military, aerospace, automobile and marine sectors.

Ship propellers are almost always made from titanium, and so are the ship's internal ballast and piping systems, and all onboard hardware exposed to seawater.

Titanium dioxide forms tiny white crystals that you see as sky-writing "ink." Skywriting typically happens at an altitude of approximately 3 km (10,000 feet) to keep the letters visible for ground-level audiences.

Some of the commercial grades of Titanium are the best choice for fishing rods, golf clubs, and other equipment.

Did You Know?

- Skywriting wasn't always whimsical. The practice dates back to World War I when pilots used it to communicate with one another and to create smokescreens to hide their movements from the enemy. Don't confuse contrails, which form as water is expelled from plane's exhaust pipes, with skywriting that is made of smoke.
- Titanium was first discovered in 1791 by William Gregor a Cornish Clergyman and amateur mineralogist while studying sand deposits in the Manaccan valley. In his sample he identified an oxide of iron and an unknown metal; he called it 'menachanite'.
- A plate of solid titanium will withstand nearly any hunting or self-defense type round of bullets indefinitely, but given enough rounds with a super high-power weapon loaded with armor-piercing bullets, it can be penetrated.

Urmi Presents Unique Lutetium

Symbol: Lu Atomic Number: 71 Atomic Mass: 175.0

Lutetium Facts

- Discovered in 1907 in Paris
- Hard, dense solid
- Good conductor of heat and electricity
- Paramagnetic
- Ductile and malleable
- Rare Earth Lanthanide Metal

Lutetium's principal commercial source is as a by-product from the processing of the rare earth phosphate mineral monazite. India, Madagascar, and South Africa have large deposits of monazite sands.

Urmi's Magical Abilities

- Urmi can shoot a laser stream from the Lutetium Aluminum Garnet at the tip of his tail that cuts through rock.

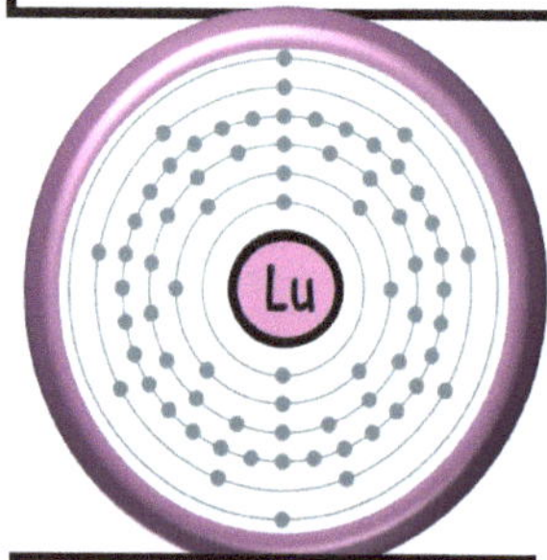

Atomic Structure

Uses For Lutetium

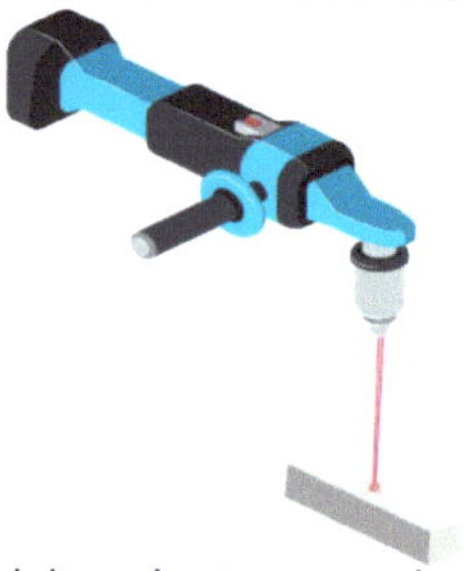

Lutetium aluminum garnet (commonly abbreviated LuAG, molecular formula Lu3Al5O12) is used in high-efficiency laser cutting devices.

LuAG is an excellent scintillation crystal suitable for thin imaging screens used in X-ray micro-radiography to create 3d images of objects without damaging them.

LuAG is also useful in the synthesis of transparent ceramics.

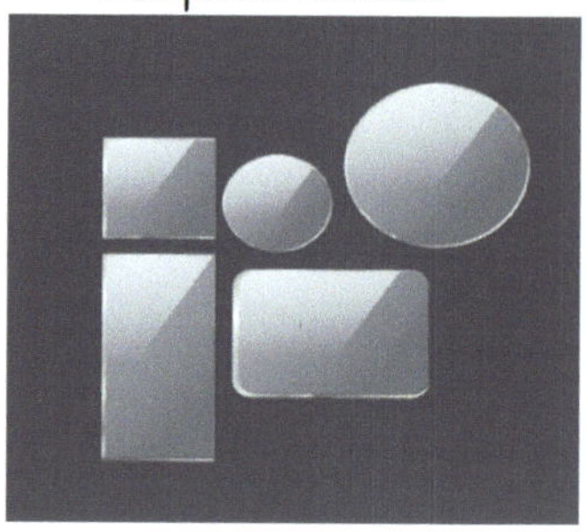

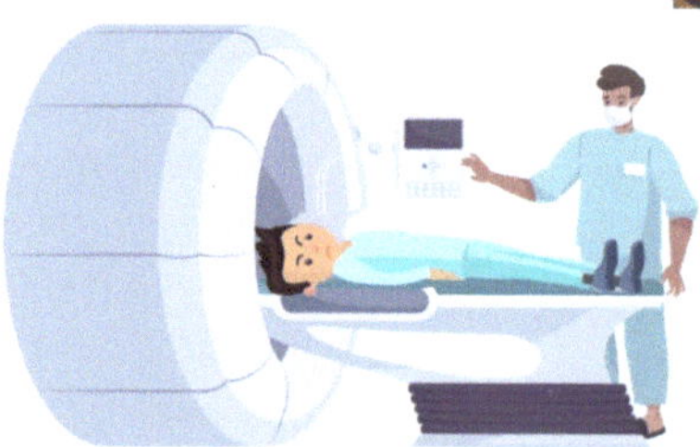

Lutetium-177 produced in nuclear reactors is used to destroy cancer cells while leaving healthy cells unaffected.

Did You Know?

- Lutetium was independently discovered by three scientists (Georges Urbain, Baron Carl Auer von Welsbach, and Charles James) as an impurity in the mineral ytterbia, which was previously thought to consist only of ytterbium. Urbain was the first to report it, in 1907, and he named the element after Lutetia, the Latin word for Paris. Lutetium was actually the last lanthanide to be isolated.
- Lutetium atoms are the smallest of any of the lanthanides. The contraction which causes that small size also causes Lutetium to have the highest melting point, the highest density, and the highest hardness of all lanthanides.
- Lutetium has its own flag based on the design of the flag of Paris, France. Rather than a sailing ship in the center of the blue and red field, it has the symbol for Alchemy. Instead of yellow fleur-de-lis there are yellow atoms. See the other differences via an internet search.

Verly Presents Vital Calcium

Symbol: Ca Atomic Number: 20 Atomic Mass: 40.08

Calcium Facts

- First isolated in 1808 in England
- Silvery white solid metal
- Good conductor of heat and electricity
- Paramagnetic
- Ductile and malleable
- Alkaline Earth Metal

Calcium constitutes 3.64 percent of Earth's crust and 8 percent of the Moon's crust. Calcium is found in many minerals, such as aragonite, gypsum and in many feldspars and zeolites. It's also in salt deposits and sea water.

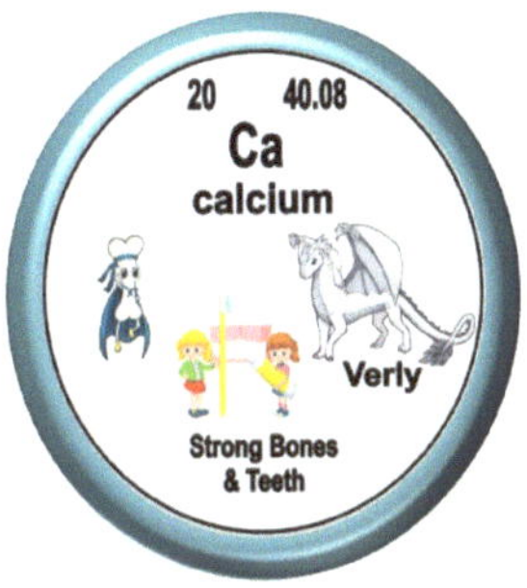

Verly's Magical Abilities

- A tap of Verly's Calcium tipped tail can heal fractures and strengthen bones in live organisms and can even strengthen inanimate structures like those made of concrete and marble.

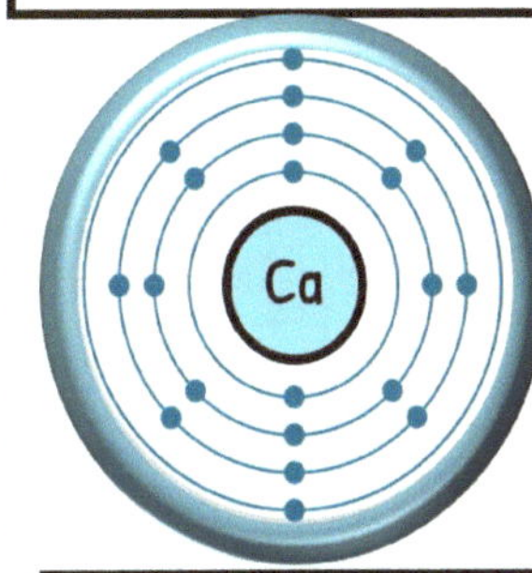

Atomic Structure

Uses For Calcium

Some foods rich in calcium include dairy products, dark green vegetables (such as spinach, purslane, chard, endive and broccoli), many herbs, legumes, nuts, fish with soft bones (such as canned sardines), fortified foods (such as plant-based alternatives to milk and yogurt). It is even in drinking water.

Calcium hydroxide solution is used as blackboard chalk.

Calcium carbonate is used to make cement and mortar used for pavements, architectural structures, foundations, motorways, roads and bridges, overpasses, parking structures and more.

Calcium carbonate is added to toothpaste. Its very gentle abrasive action gently polishes tooth enamel without damaging it.

Did You Know?

- The name calcium comes from the Latin word calx which means lime. Ancient civilizations originally obtained calcium oxide by heating limestone. The lime-based mortars were used for building purposes.
- Almost all organisms rely on calcium. It is the most abundant metallic element in the human body, found mainly in bones and teeth. The amount of calcium required for proper growth in many creatures varies dramatically across life stages.
- Currently, the largest amount of calcium is used by steel industries for making steel. Calcium helps make the steel better by making it easier to shape and keeping it pure and strong.

White Wing Presents White Gold

An alloy of Gold, Copper, Silver, Zinc and Nickel

White Gold Facts

- First commercially available in 1912 in Germany
- Very malleable solid
- and electricity
- Not naturally magnetic
- Ductile or malleable
- Transition Metal Alloy

White Gold is comprised of a mixture of several transition metals. 14k (karat) contains Gold, Copper, Zinc and Nickel. The addition of Silver makes it 18k. 18k is 75% pure and 14k is 58% pure gold. 24k is 100% pure.

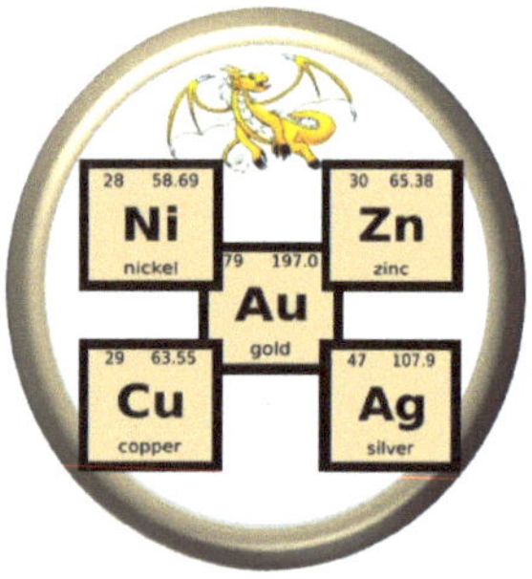

White Wing's Magical Abilities

- He can spin thin golden wire from the tip of his White Gold tipped tail. It can be used for jewelry and other industrial purposes.

Alloy Percentages

Uses For White Gold

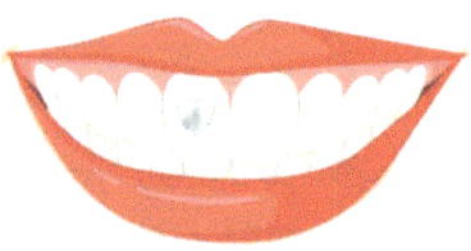

How about some White Gold jewelry? You can even get some for your teeth.

Gold is known to have been used in dentistry as early as 700 B.C. Etruscan "dentists" used gold wire to fasten replacement teeth into the mouths of their patients. White Gold is still used in braces today.

Gold alloys are used in connectors, switch and relay contacts, soldered joints, connecting wires and connection strips. You will find gold alloys in phones, calculators and most large electronic appliances – even in standard desktop or laptop computers and televisions.

Alloyed gold coins are typically 91.7% gold by weight, with the remainder being silver and copper.

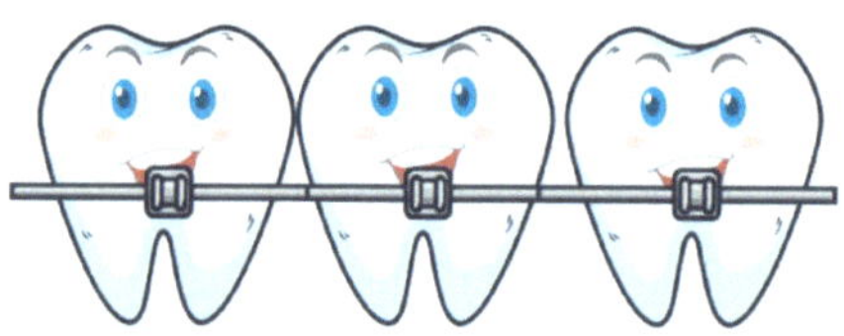

Did You Know?

- Although the origin of white colored gold dates back to 1710, where it was discovered by two alchemists in Germany, it wasn't used widely until much later. White gold as we know it today was really popularized in the 1920s. Although it might seem strange, this luxury material came to popularity in wartime. And it actually came about because of the popularity, utility, and scarcity of Platinum.
- White Gold metal naturally has a yellowish shade so most of the time it is coated with a thin layer of Rhodium or Palladium to give it a depth of color and shine. The white metal coating is what makes this alloy look more recognizably like how people think it should look.
- Rhodium plating was also added as a protective layer between the Nickel in White Gold and the wearer who may be allergic to it.

Xoe Presents Xanthous Sulfur

Symbol: S Atomic Number: 16 Atomic Mass: 32.06

Sulfur Facts

- Recognized as an element in 1777 by Antoine Lavoisier
- Brittle, crystalline solid
- Weak conductor of heat and electricity
- Antiferromagnetic
- Neither ductile or malleable
- Non-metal

Pure Xanthous (Yellow) Sulfur is deposited near hot springs and volcanoes but the main source of commercial Sulfur is through recovery from crude oil during the refining process.

Xoe's Magical Abilities

- When Xoe scratches the tip of her Xanthous (yellow) Sulfur tail against a rough surface she can spark a small, or raging fire.

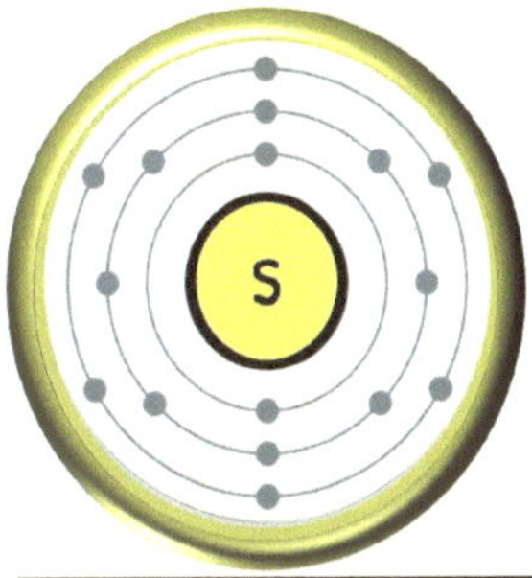

Atomic Structure

Uses For Sulfur

Match heads contain an oxidising agent, abrasive materials and other additive compounds such as sulfur, added as fuel to help the match head burn.

Sulfur is an important element in sulfa-drugs that are used for bronchitis and coughs. This medicine helps kill the bacteria making you sick.

Coffee contains several sulphuric compounds that contribute to its signature aroma and taste. Different roasting and processing methods can induce different reactions in the sulfur content, producing different tastes, odors, and aromas.

Sulfur is used in pesticides and fertilizers throughout the world.

Did You Know?

- In ancient times, sulfur was called brimstone. It has been in use since before 2000 BCE. Pre-Roman civilizations used burned brimstone as a medicine and used "bricks" of sulfur as fumigants, bleaching agents, and incense in religious rites.
- Natural sulfur is very yellow (xanthous) colored. If you melt sulfur, you get a blood-red liquid. If you set it on fire, you get a blue flame.
- Surprisingly, pure sulfur has no smell and no taste! But did you know, sulfur compounds called mercaptans give skunks their defensive odor; and rotten eggs and stink bombs get their distinctive aroma because of hydrogen sulfide. Also, the main reason that intestinal gas has that nasty odor is that the large infestine is full of bacteria that release sulfur compounds as waste.

Yago Presents Yttrium

Symbol: Y Atomic Number: 39 Atomic Mass: 88.91

Yttrium Facts

- Discovered 1787 in Sweden
- Brittle, crystalline solid
- Good conductor of heat and electricity
- Paramagnetic
- Malleable and Ductile
- Rare Earth Transition Metal

Yttrium is found with other rare earth mineral ores, like monazite and bastnaesite, in Canada, Brazil, Australia, Hungary, and Greece.

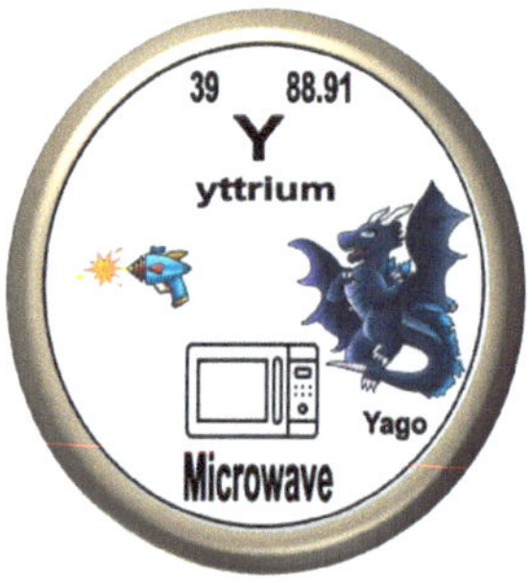

Yago's Magical Abilities

- His Yttrium-aluminium garnet (YAG) tipped tail can cut through metal like a laser.

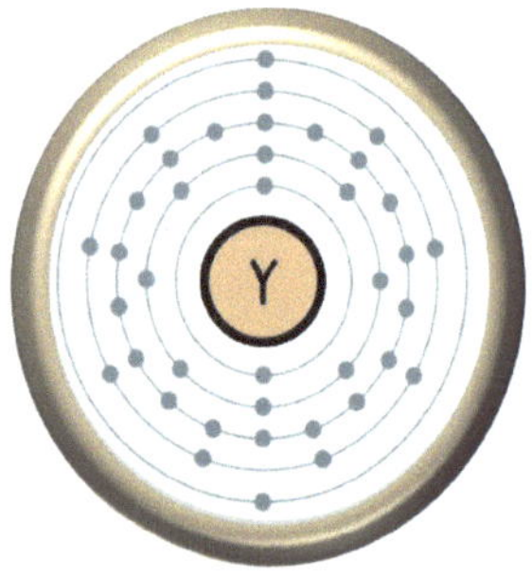

Atomic Structure

Uses For Ytterium

Yttrium is used in the making of microwave filters for radar equipment like those that track air traffic.

Yttrium Aluminum Garnet (YAG) is a synthetic laser crystal widely used as active laser media in solid-state lasers in medicine and manufacturing.

Yttria stabilized zirconium oxide is used in high temperature applications, such as in thermal plasma sprays to protect aerospace high temperature surfaces.

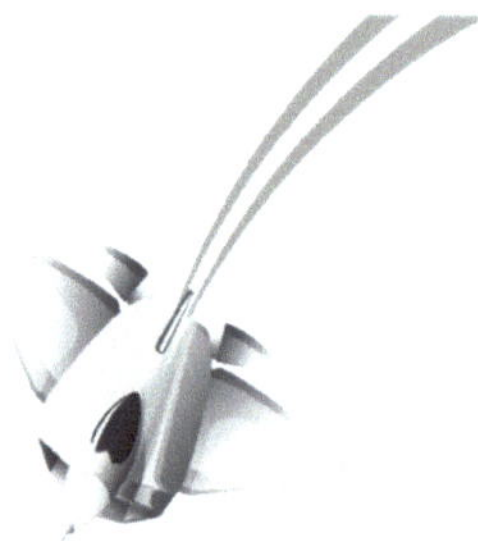

Did You Know?

- Synthetic garnets made with an yttrium-aluminum composite were commonly sold in the 1970s as simulated diamonds and other gemstones, but they eventually gave way to the cubic zirconia.
- The radioactive isotope yttrium-90 is used in radiation therapy to treat liver cancer and some other cancers.
- Yttrium (pronounced as IT-ree-em), contains 4 oxides— yttrium oxide (white), terbium oxide (yellow), rose-colored erbium oxide and ytterbium oxide (white). An oxide is a chemical compound containing at least one oxygen atom and one other element in its chemical formula.

Zora Presents Zirconium

Symbol: Zr Atomic Number: 40 Atomic Mass: 91.22

Zirconium Facts

- Used in jewelry in ancient times. Officially discovered in 1789.
- Hard, brittle solid
- Conductor of heat and electricity
- Paramagnetic
- Ductile and malleable
- Transition Metal

Zirconium naturally occurs in the raw gemstone Zircon, which are mined in the United States, Australia, Brazil, South Africa, Russia, and Sri Lanka.

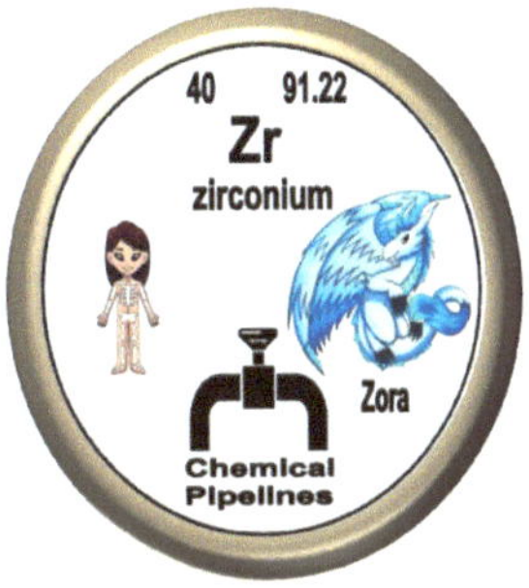

Zora's Magical Abilities

- She can magically strengthen bones, joints and teeth with a touch of her Zirconium tipped tail.

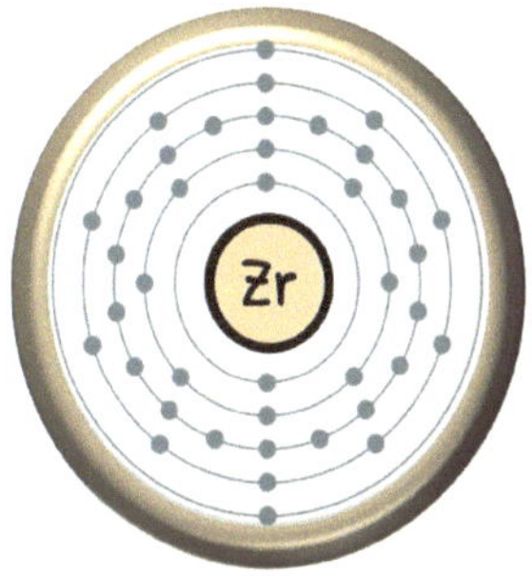

Atomic Structure

Uses For Zirconium

Zircon sand is used for heat-resistant linings for furnaces and giant ladles for molten metal; and to make foundry molds.

Zirconium is so strong that even scissors and knives can be made from it.

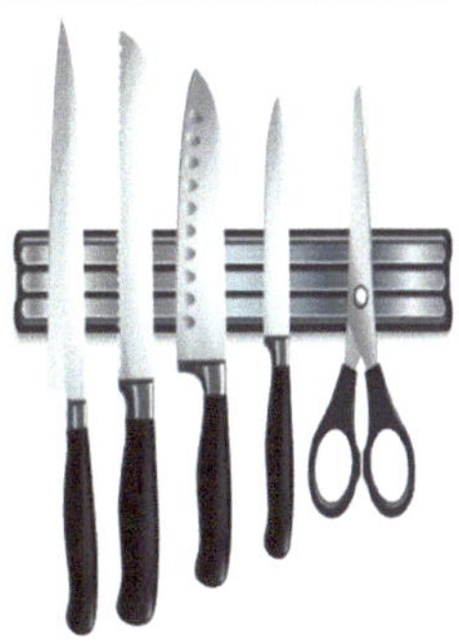

Zirconium is utilized in a wide range of medical procedures, including the creation of dental crowns and bridges, as well as implants for artificial joints.

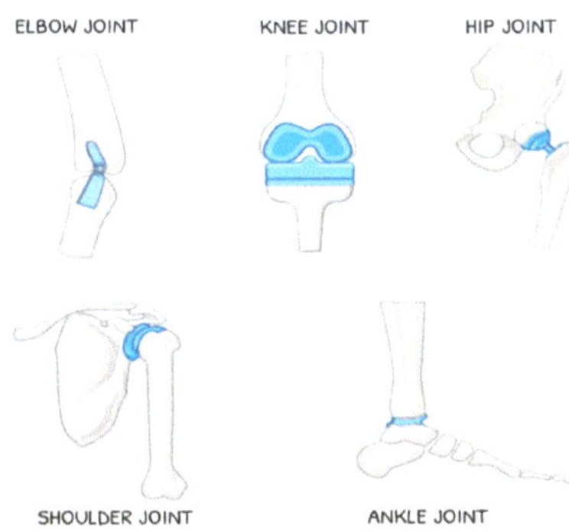

When alloyed with niobium, zirconium is superconductive at low temperatures and is used to make superconducting magnets such as those used for maglev trains.

Did You Know?

- It is estimated that up to 90 per cent of Zirconium produced in the world is used for nuclear power.
- It is known that Zirconia is a bone-biocompatible material so Zirconia fillers are added to traditional PMMA bone cement for preventing potential post-surgical infection.
- Zirconium is a derivative of the mineral zircon, the oldest mineral on Earth, dating back more than 4.4 billion years! The name 'zirconium' is also related to the Persian Arabic word 'zargun', meaning gold-like.

Magical Elements of The Periodic Table

Magical elementals from the Magical Elements of the Periodic Table books present all of the elements of the periodic table in fantastical and real life terms.

In the books, each elemental character has magical powers based on the properties of the elements that come from the land, air and water. They are the perfect group to introduce you to metals, metalloids, non-metals, halogens, noble gases and much more.

Unicorns, dragons, alchemists, knights, and goblins will show you how people of this world always have and always will depend upon the elements that our earth provides for all of our needs.

Remember, "No Metal— No Magic." . . .And no technology.

No Metal

Actinium To Zirconium

No Magic

Use this Periodic Table as you would any other to spark an interest in the magical and real world properties of all the elements known today. You may be surprised at how prominently they feature in our every day lives.

	1	2	3	4	5	6	7	8	9	10	11	12	13	14	15	16	17	18
1	1 1.008 **H** hydrogen Hildy Textile Manufacturing																	2 4.003 **He** helium Hetha Balloons
2	3 6.94 **Li** lithium Lillian Batteries	4 9.012 **Be** beryllium Berwyn Musical Instrument											5 10.81 **B** boron Boroleas Sports Equipment	6 12.01 **C** carbon Cole Charcoal	7 14.01 **N** nitrogen Nitra Food Packaging	8 16.00 **O** oxygen Ozzy Air	9 19.00 **F** Fluorine Fleure Strong Bones and Teeth	10 20.18 **Ne** neon Jalan Advertising Signs
3	11 22.99 **Na** sodium Sorn Salt	12 24.31 **Mg** Magnesium Maggie In Your Bones											13 26.98 **Al** aluminium Alumna Airplanes	14 28.09 **Si** silicon Silonar Glass	15 30.97 **P** phosphorus Phova Fertilizer	16 32.06 **S** sulfur Xoe Matches	17 35.45 **Cl** chlorine Krystix Swimming Pools	18 39.95 **Ar** argon Areg Light Bulbs
4	19 39.10 **K** Potassium Pearl Saline Drips	20 40.08 **Ca** calcium Verly Teeth	21 44.96 **Sc** scandium Scandra Bicycles	22 47.87 **Ti** titanium Tilly Aerospace	23 50.94 **V** vanadium Vana Black Printer Ink	24 52.00 **Cr** chromium Crowmist Stainless Steel	25 54.94 **Mn** manganese Mangar Earth Movers	26 55.85 **Fe** Iron Iown Bicycle Chains	27 58.93 **Co** cobalt Coriss Magnets	28 58.69 **Ni** nickel Nix Guitar Strings	29 63.55 **Cu** copper Cuprum Money	30 65.38 **Zn** Zinc Dr. Zinko Suntan Lotion	31 69.72 **Ga** gallium Gallant LED Displays	32 72.63 **Ge** germanium Gemel Camera Lense	33 74.92 **As** arsenic Arkyn Poison	34 78.97 **Se** selenium Selenice Printers	35 79.90 **Br** bromine Brogach Photography Film	36 83.80 **Kr** krypton Krypto Detect Leaks
5	37 85.47 **Rb** Rubidium Ruby Night Vision Glasses	38 87.62 **Sr** strontium Strauna Computer Screens	39 88.91 **Y** yttrium Yago Microwave	40 91.22 **Zr** zirconium Zora Chemical Pipelines	41 92.91 **Nb** niobium Nonnach Mag Lev Trains	42 95.95 **Mo** molybdenum Maximo Cutting Tools	43 98 **Tc** technetium Tephen Radio Active Diagnosis	44 101.1 **Ru** ruthenium Ruth Electrical Switches	45 102.9 **Rh** rhodium Rovana Finish for Jewelry	46 106.4 **Pd** palladium Paedin Concert Flute	47 107.9 **Ag** Silver Silubhra Ventilator	48 112.4 **Cd** cadmium Cadmus Power Tools	49 114.8 **In** indium Iker Liquid Crystal Display (LCD)	50 118.7 **Sn** Tin Tinam Liquid Crystal Display	51 121.8 **Sb** antimony Antz Flame Resistant Fabric	52 127.6 **Te** tellurium Tellan Vulcanized Rubber	53 126.9 **I** Iodine Jody (Iodium) Cloud Seeding	54 131.3 **Xe** xenon Xena Used To Catch Speeders

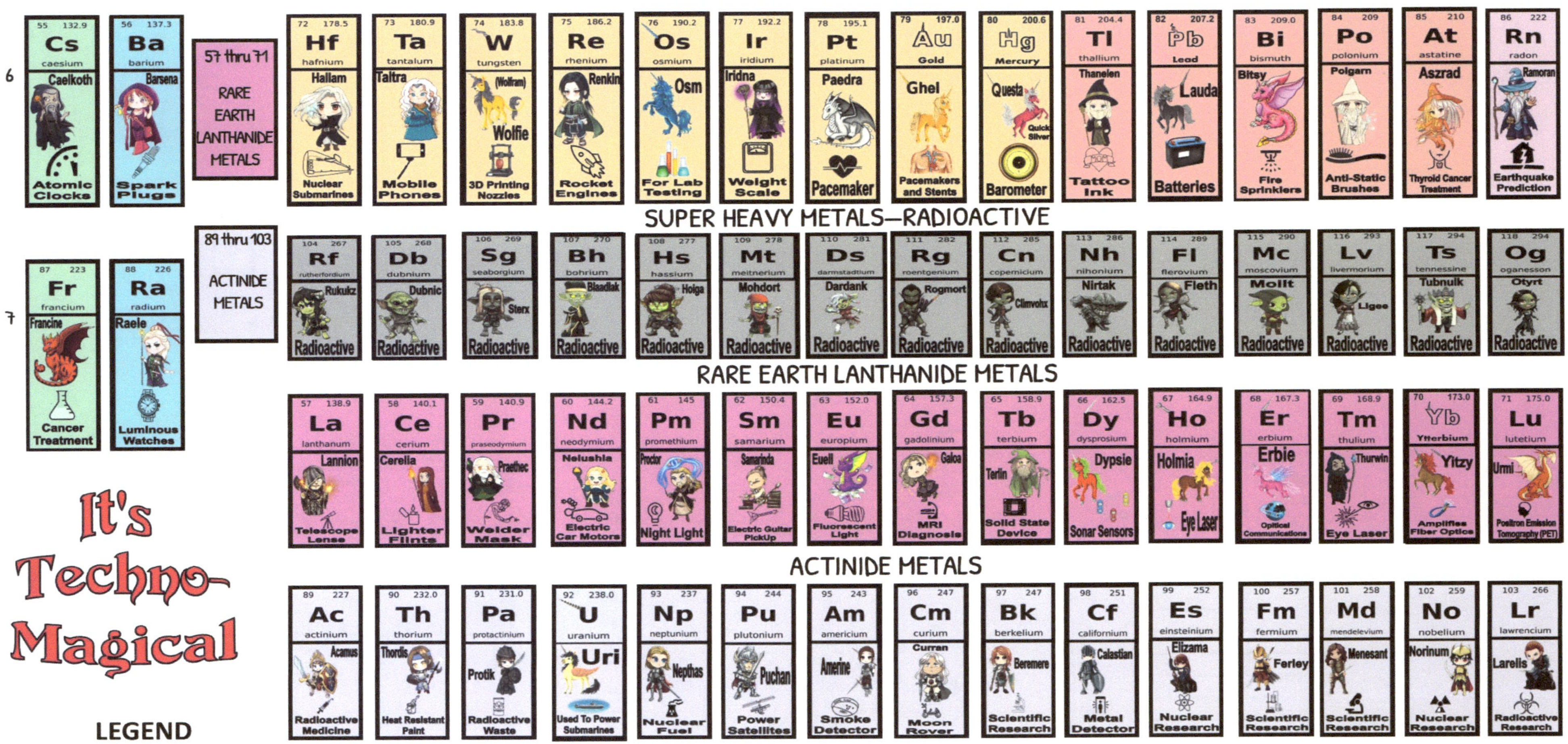

It's Techno-Magical

LEGEND

- Alkali Metals
- Alkali Earth Metals
- Transition Metals
- Post-Transition (or Other Metals)
- Metalloids
- Non-Metals
- Halogens
- Noble Gases
- Rare Earth Lanthanide Metals
- Actinide Metals
- Super Heavy—Radioactive

Alloys are created when 2 or more metals are combined. Compounds are created when 2 or more non-metals are combined.

EXAMPLE OF A COMPOUND

Quincy

Quick Lime =

Used for Concrete

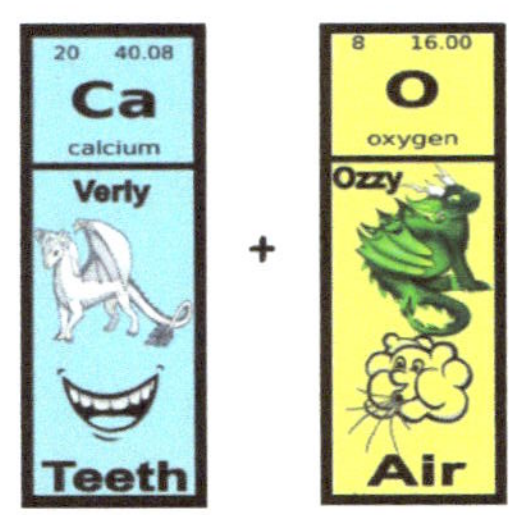

Both Carbon and Oxygen are reactive nonmetals.

EXAMPLE OF AN ALLOY

White Wing

Used for jewelry, dental amalgams plus connectors, and switch and relay contacts for electronics.

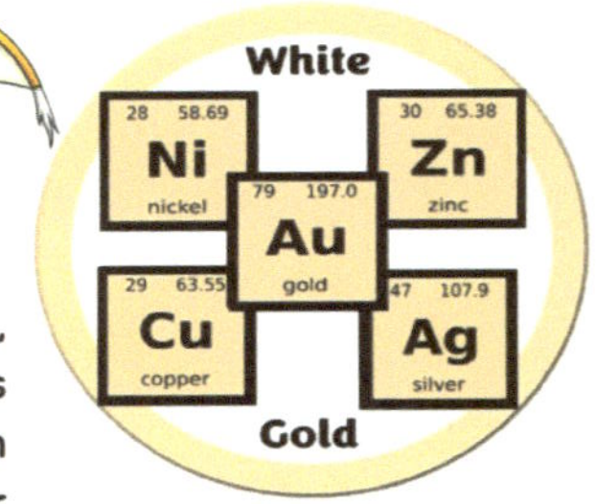

White Gold

Includes 58.5 % gold, 22% copper, 8% zinc, 7% nickel, 4.5% silver and possibly other elements.

Sybrina.com

Structure Of Elements In The Periodic Table

Periodic tables are laid out in rows and columns.

Vertical columns are called Groups.

Each element is placed in a specific location because of its atomic structure. Elements are arranged in Families.

Horizontal rows are called Periods.

All rows read left to right.

The last two periods are part of periods 6 & 7.

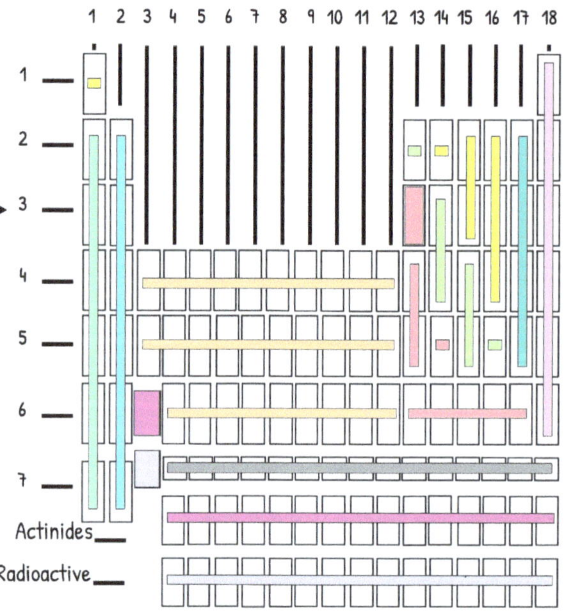

The Families are:

- Alkali Metals
- Alkali Earth Metals
- Transition Metals
- Post-Transition (or Other Metals)
- Metalloids
- Non-Metals
- Halogens
- Noble Gases
- Rare Earth Lanthanide Metals
- Actinide Metals
- Heavy-Radioactive

The term 'Element' is used to describe atoms with specific characteristics.
Every element in the first column or Group has 1 electron in the outer orbital (shell).
Every element in the second column (group two) has two electrons in the element's outer orbital.
The number designation of each Group represents the number of electrons in the element's outer orbital—except for Group 18, Period 1—Helium. It only has 2 electrons.
Those electrons, called Valence Electrons, are what chemically bond with other elements.

Atomic Structure of Element: The atomic structure of an element refers to the arrangement of protons and neutrons in the nucleus of the atom, and the electrons in the electron cloud around the nucleus. Group 1, Period 1—Hydrogen is the only element that has no neutrons.

Sb—Antimony
Group 15—Period 5

Atomic Number— Indicates the number of protons in the nucleus and the number of electrons in the atom (Element).

Element Symbol— The identifying letters for the element are an abbreviation of the element name. Only one or two letters are used for element symbols. Sb = Antimony

The Elemental Dragon introducing this Element in the Magical Elements of the Periodic Table Book 2 is Antz the Dragon With The Antimony Tipped Tail.

Atomic Mass (or Weight) — Equals the number of Protons + Neutrons in the Element's nucleus.

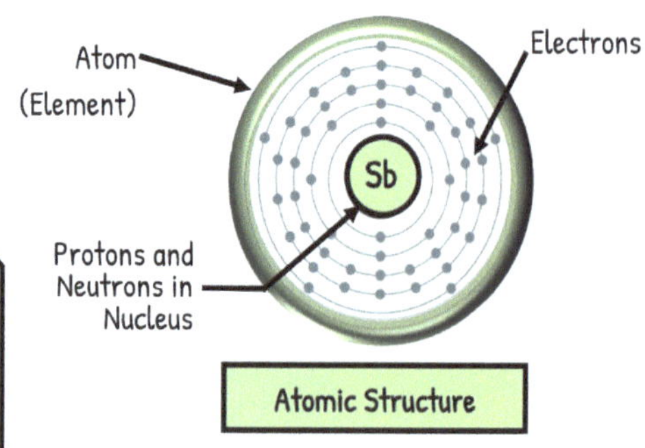

Element Name—
The name of this element is Antimony. It is a Metalloid.

An every day use for the element.
IE: Antimony is used to make flame resistant fabric.

Types of Elements On The Periodic Table

Alkali Metals—Some metals on the periodic table are soft and shiny. They are so soft that they can be cut with a knife! These metals are excited to give away electrons to elements in need, making them highly reactive. This electron transfer creates a compound known as a salt. Surprisingly, these metals are not found in nature alone; they must be extracted from other sources. Examples of these metals include lithium, sodium, potassium, rubidium, cesium, and francium.

Alkali Earth Metals—The elements in column 2 of the periodic table have 2 outer electrons in their shell. This makes them very active with nonmetals that need electrons to stay stable. When they react, they make something called a salt. They are often found in nature all by themselves, and they can even conduct electricity. The elements are beryllium, magnesium, calcium, strontium, barium, and radium.

Post-Transition (or other Metals)— Elements directly to the right of the transition metals. They are known as "poor metals: and are soft and brittle. These include aluminum, gallium, indium, tin, thallium, lead, bismuth, zinc, cadmium and mercury.

Transition Metal—The main metals are found in the middle and bottom rows of the periodic table. They look like metal, can conduct electricity, can bend and be shaped easily. The period 4 transition metals are scandium, titanium, vanadium, chromium, manganese, iron, cobalt, nickel, copper, and zinc. The period 5 transition metals are yttrium, zirconium, niobium, molybdenum, technetium, ruthenium, rhodium, palladium, silver, and cadmium. The period 6 transition metals are lanthanum, hafnium, tantalum, tungsten, rhenium, osmium, iridium, platinum, gold, and mercury. The period 7 transition metals are the naturally-occurring actinium, and the artificially produced elements rutherfordium, dubnium, seaborgium, bohrium, hassium, meitnerium, darmstadtium, and roentgenium.

Metalloids—The elements called metalloids are a mix of metals and nonmetals. They look like metals, but can't conduct electricity very well. They also break easily and act like nonmetals. These include boron, silicon, germanium, arsenic, antimony, tellurium, astatine, and polonium.

Non-Metals—These elements reside in columns 15-17, and can be gases, liquids, or solids. They don't conduct heat or electricity. The solids are brittle, and they have no metallic luster. They readily accept electrons from metals to form salts. These include nitrogen, oxygen, fluorine, chlorine, bromine, and iodine.

Halogens—Halogen chemicals are a special type of element. When they mix with metal, they become a kind of salt. Halogens are super reactive because they like to take an electron from metals. They can be found in column 17 of the element table. Some of them can be found in nature, but most are very dangerous and can hurt you if you touch them. They include fluorine, chlorine, bromine, iodine, and the radioactive elements astatine and tennessine.

Noble Gases—These elements reside in column 8. They are all odorless, colorless gases that are chemically very stable (inert). They don't generally form compounds by bonding with another element. These include helium, neon, argon, krypton, xenon, and radon.

Lanthanide Rare Earth Minerals—The Japanese call them "the seeds of technology." The US Department of Energy calls them "technology metals." These elements have atomic numbers 57-71. They are vital to industry. They can be added to metals to strengthen them to make alloys such as stainless steel, used to refine crude oil, and are crucial in producing technology—electronics, telecommunications, and metal devices to name a few. They are lanthanum, cerium, praseodymium, neodymium, promethium, samarium, europium, gadolinium, terbium, dysprosium, holmium, erbium, thulium, ytterbium, lutetium, scandium and yttrium.

Actinide Metals—Any of a series of chemically similar metallic elements with atomic numbers ranging from 89 (actinium) to 103 (lawrencium). All of these elements are radioactive, and two of the elements, uranium and plutonium, are used to generate nuclear energy. The lanthanides and actinides are sometimes called the inner transition metals, referring to their properties and position on the table. They are actinium, thorium, protactinium, uranium, neptunium, plutonium, americium, curium, berkelium, californium, einsteinium, fermium, mendelevium, nobelium, and lawrencium.

Super Heavy—Radioactive—Superheavy elements are those elements with a large number of protons in their nucleus. Elements with more than 92 protons are unstable; they decay to lighter nuclei with a characteristic half-life. They do not occur in large quantities (if at all) naturally on earth, and only exist briefly under highly controlled circumstances. They include lawrencium, rutherfordium, dubnium, seaborgium, bohrium, hassium, meitnerium, darmstadtium, roentgenium, copernicium, nihonium, flerovium, moscovium, livermorium, tennessine, and oganesson.

Alloys

An **alloy** is a mixture of chemical elements of which at least one is a metal. An alloy is a solid. Unlike chemical compounds with metallic bases, an alloy will retain all the properties of a metal in the resulting material, such as electrical conductivity, ductility, opacity, and luster, but may have properties that differ from those of the pure metals, such as increased strength or hardness. In some cases, an alloy may reduce the overall cost of the material while preserving important properties. In other cases, the mixture imparts synergistic properties to the constituent metal elements such as corrosion resistance or mechanical strength. Some of the most common alloys are

BeCu =

Beryllium Copper alloy is used for strengthening tools, musical instruments, and sports equipment

0.5% min / 3% max

+

96% min / 97% max

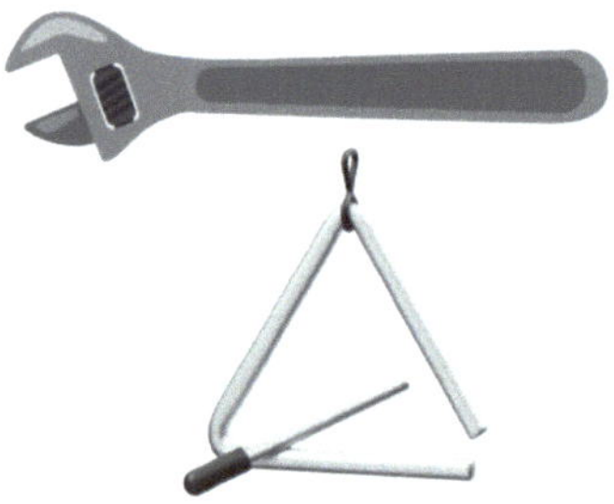

Brass =

Used for decoration, plumbing, instruments

65% min / 90% max

+

10% min / 35% max

+

May also Include iron, lead, manganese, aluminum, silicon and other elements.

Steel =

Used for structures, cutlery, car bodies, rails

50% min / 99% max

+

0.1% min / 2.5% max

+

May also Include manganese, silicon, copper, nitrogen, niobium, titanium, or sulfur

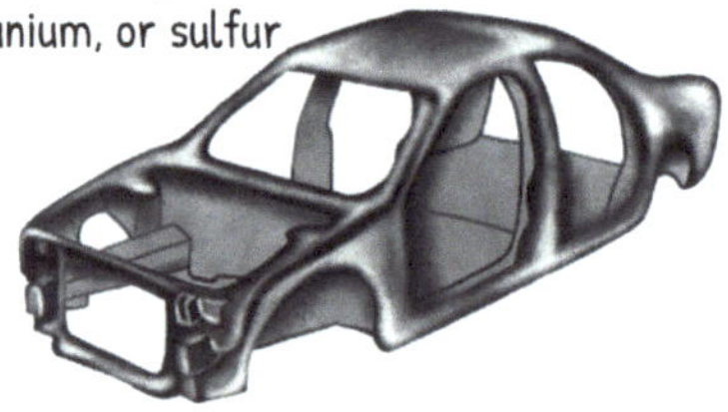

White Gold =

18K—Used for jewelry & orthodontics

58.5%

+

22%

+

7%

+

8%

+

4.5%

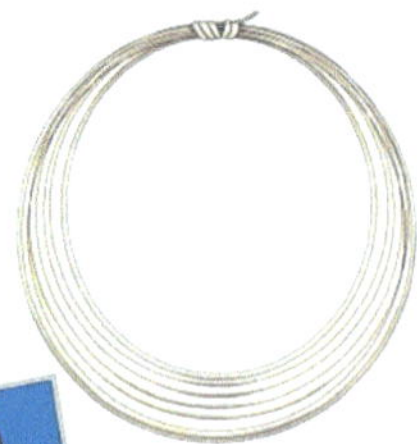

Some other common alloys are Bronze, Cast Iron, Cupronickel, Magnalium, Mischmetal, Nichrome, Nitinol, Pewter, Solder, Sterling Silver and Tungsten Carbide.

The above chart only shows a few of the hundreds of metal combinations. For instance, 24 carat gold is a pure naturally occurring yellow metal. There are four basic shades of gold alloys: yellow gold, white gold, rose gold, and green gold. A huge range of other colored golds are also possible, including red (gold and copper), grey (gold, iron and copper), purple (gold and aluminum), blue (gold and iron) and black (gold and cobalt), depending on the amounts of different metals alloyed together.

Compounds

A **compound** is a substance formed due to the chemical union (a chemical reaction) between two or more atoms or molecules. Ideally there should be two or more elements to form a compound. Most of the compounds are from a non-metallic origin.

Quicklime =

Used for– building concrete structures, cleaning water and making sugar.

+

Both Carbon and Oxygen are reactive nonmetals.

Sodium Fluoride =

Used for–the fluoridation of drinking water, in toothpaste, in metallurgy, and as a flux, and is also used in pesticides and rat poison.

+

Sodium Fluoride is a simple ionic compound, made of the sodium (Na^+) cation and fluoride (F^-), an anion of Flourine.

Potassium Iodide =

It's medical use is to block absorption of radioactive iodine by the thyroid gland.

+

Potassium iodide (also called KI) is a salt of stable (not radioactive) iodine. Potassium is an alkali metal and Iodine is a reactive nonmetal.

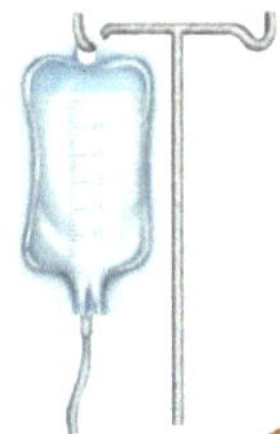

Calcium Bromate =

It is used as a bread dough and flour "improver" or conditioner

+

Calcium is an alkaline earth metal and Bromine is a reactive nonmetal.

Can you guess the most important compound of all?

+

= ??????

(Answer can be found below.)

The above chart only shows a few of the millions of combinations of elements that create compounds.

Iodine, alone has 37 compounds. Some of them are Ammonium iodide (NH_4I), Cesium iodide (CsI), Copper(I) iodide (CuI), Hydroiodic acid (HI), Iodic acid (HIO_3), Iodine cyanide (ICN), Iodine heptafluoride (IF_7), Iodine pentafluoride (IF_5), Lead(II) iodide (PbI_2), Lithium iodide (LiI), Nitrogen triiodide (NI_3), Potassium iodate (KIO_3), Potassium iodide (KI), Sodium iodate ($NaIO_3$), and Sodium iodide (NaI).

Hydrogen is used in more compounds (nearly 100) than any other element because it can form bonds with almost all metals, metalloids, and non-metals. Some of the most common are water (H_2O), Hydrogen Peroxide (H_2O_2) and table sugar ($C_{12}H_{22}O_{11}$).

Neon forms no known compounds.

All Of The Periodic Table Elements Listed Alphabetically

Elements Listed In Red Are Featured In This Book

ACTINIUM–*AC*–89

ALUMINUM–*AL*–13

AMERICIUM–*AM*–95

ANTIMONY–*SB*–51

ARGON–*AR*–18

ARSENIC–*AS*–33

ASTATINE–*AT*–85

BARIUM–*BA*–56

BERKELIUM–*BK*- 97

BERYLLIUM–*BE*–4

BISMUTH–*BI*–83

BOHRIUM–*BH*–107

BORON–*B*–5

BROMINE–*BR*–35

CADMIUM–*CD*–48

CALCIUM (Vital)–CA–20

CALIFORNIUM–*CF*–98

CARBON–*C*–6

CERIUM–*CE*–58

CESIUM–*CS*–55

CHLORINE (Keen)–*CL*–17

CHROMIUM–*CR*–24

COBALT–*CO*–27

COPERNICIUM–*CN*–112

COPPER–*CU*–29

CURIUM–*CM*–96

DARMSTADTIUM–*DS*–110

DUBNIUM–*DB*–105

DYSPROSIUM–*DY*–66

ERBIUM–*ER*–68

EINSTEINIUM–*ES*–99

EUROPIUM–*EU*–63

FERMIUM–*FM*–100

FLEROVIUM–*FL*–114

FLUORINE–*F*–9

FRANCIUM–*FR*–87

GADOLINIUM–*GD*–64

GALLIUM–*GA*–31

GERMANIUM–*GE*–32

GOLD–*AU*–79

HAFNIUM–*HF*–72

HASSIUM–*HS*–108

HELIUM–*HE*–2

HOLMIUM–*HO*–67

HYDROGEN–*H*–1

INDIUM–*IN*–49

IODINE *(JODIUM)* – *I*–53

IRIDIUM–*IR*–77

IRON–*FE*–26

KRYPTON–*KR*–36

LANTHANUM–*LA*–57

LAWRENCIUM–*LR*–103

LEAD–*PB*–82

LITHIUM–*LI*–3

LIVERMORIUM–*LV*–116

LUTETIUM (Unique)–*LU*–71

MAGNESIUM–*MG*–12

MANGANESE–*MN*–25

MEITNERIUM–*MT*–109

MENDELEVIUM–*MD*–101

MERCURY *(QUICK SILVER)* – *HG*–80

MOLYBDENUM–*MO*–42

MOSCOVIUM–*MC*–115

NEODYMIUM–*ND*–60

NEON (Jazzy)–*NE*–10

NEPTUNIUM–*NP*–93

NICKEL–*NI*–28

NIHONIUM–*NH*–113

NIOBIUM–*NB*–41

NITROGEN–*N*–7

NOBELIUM–*NO*–102

OGANESSON–*OG*–118

OSMIUM–*OS*–76

OXYGEN–*O*–8

PALLADIUM–*PD*–46

PHOSPHORUS–*P*–15

PLATINUM–*PT*–78

PLUTONIUM–*PU*–94

POLONIUM–*PO*–84

POTASSIUM–*K*–19

PRASEODYMIUM–*PR*–59

PROMETHIUM–*PM*–61

PROTACTINIUM–*PA*–91

RADIUM–*RA*–88

RADON–*RN*–86

RHENIUM–*RE*–75

RHODIUM–*RH*–45

ROENTGENIUM–*RG*–111

RUBIDIUM–*RB*–37

RUTHENIUM–*RU*–44

RUTHERFORDIUM–*RF*–104

SAMARIUM–*SM*–62

SCANDIUM–*SC*–21

SEABORGIUM–*SG*–106

SELENIUM–*SE*–34

SILICON–*SI*–14

SILVER–*AG*–47

SODIUM–*NA*–11

STRONTIUM–*SR*–38

SULFUR (Xanthous)–*S*–16

TANTALUM–*TA*–73

TECHNETIUM–*TC*–43

TELLURIUM–*TE*–52

TENNESSINE–*TS*–117

TERBIUM–*TB*–65

THALLIUM–*TL*–81

THORIUM–*TH*–90

THULIUM–*TM*–69

TIN–*SN*–50

TITANIUM–*TI*–22

TUNGSTEN–*W (WOLFRAM)* – 74

URANIUM–*U*–92

VANADIUM–*V*–23

XENON–*XE*–54

YTTERBIUM–*YB*–70

YTTRIUM–*Y*–39

ZINC–*ZN*–30

ZIRCONIUM–*ZR*–40

Definitions

Atomic Structure of Element: The atomic structure of an element refers to the arrangement of protons and neutrons in the nucleus of the atom, and the electrons in the electron cloud around the nucleus.

Atomic Number: An element's atomic number refers to the number of protons it has in its nucleus. In a neutral atom the number of protons always equals the number of electrons.

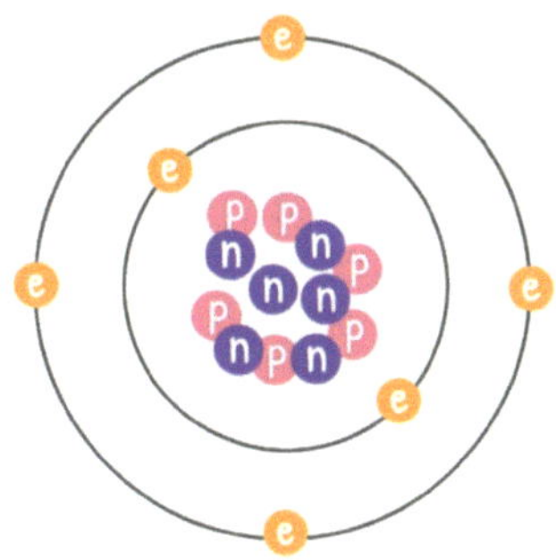

Carbon atom

Atomic Weight (Mass) of Element: The atomic mass of an element is how heavy it is. It is made up of protons and neutrons that are in the middle of the element. Some elements have different versions with different amounts of neutrons, but they still have the same amount of protons. The atomic mass is the average weight of all these versions of the element.

Allotrope: Allotropes are different forms of an element that look and act different, but are made of the same stuff. Some elements have more than one form. For instance, carbon can be a shiny diamond or a gray pencil lead called graphite.

Isotope: Isotopes are different types of atoms that have the same parts, like protons and electrons, but they have a different number of neutrons. For example, the three most stable isotopes of hydrogen: protium (A = 1), deuterium (A = 2), and tritium (A = 3).

Crystalline Structure of Element: The crystalline structure of an element is how its atoms, ions, or molecules stick together in a pattern to make a cool crystal shape.

Ferrous and Non-Ferrous Metals: When we say ferrous metal, it means that iron is a big part of the metal. But if there's only a little bit of iron in the metal, we call it non-ferrous. The word "ferrous" comes from Latin and means iron, which is why iron's symbol is Fe.

Ductile Metals: These are capable of being made into long, thin wire or thread. Copper and Silver are ductile metals.

Malleable Metals: These can be hammered or rolled into thin sheets without cracking or breaking. Gold is malleable.

Ferromagnetic: Materials that are strongly attracted to a magnet. Such materials can be permanently magnetized. These include the elements iron, nickel and cobalt and their alloys, some alloys of rare-earth metals, and some naturally occurring minerals such as lodestone.

Magnetostriction—This is the term for a special thing that happens to magnetic materials. When these materials get turned into magnets, they also change their shape or size.

Paramagnetic: Slightly attracted to a magnetic field, but do not retain magnetic properties once the field is removed.

Diamagnetic: Slightly repelled by a magnetic field, but do not retain magnetic properties once the field is removed.

Electrical Properties: Conductor—a thing that lets electricity flow through it. Semi-conductor—a special material (usually silicon) that can conduct electricity, but not as well as metal. Insulator (non-conductor)—a material (usually glass) that stops electricity from flowing.

Reactive Gas: These gases are really good at reacting with stuff! They are called "sticky gases" because they can react to things like plastic and wet surfaces when they touch them. These are nitrogen, oxygen, hydrogen, carbon dioxide, fluorine, and chlorine.

Non-Reactive Gas: An inert gas is like a super shy gas that doesn't like to hang out with other chemicals. It doesn't make any new friends by reacting with them, so it doesn't form any chemical compounds. We also call these special gases "noble gases."

Dear Reader

I hope Magical Elements of The Periodic Table Presented Alphabetically By The Elemental Dragons has helped you learn a little something about the magic of metals and other elements from the periodic table.

Most of the dragons were illustrated by Pranavva (@pranavva_ on Instagram & @nearwasa on Fiverr) plus a few by Tamagon (@tamagonstudio on Instagram & Fiverr). The everyday items were illustrated by Usman (@usman_arts5 on Fiverr).

This book is a dragon "alphabet book". The future books do not cover the alphabet. Next up is a wizard periodic table book with many more interesting characters and elements. This will eventually be a 6 volume set featuring wizards, knights and goblins introducing all 118 elements. Every one of them is amazing and very necessary to our everyday lives. All of the elements are truly

Techno-magical.

A lot of research went into every page of this book as well as all the other Magical Elements of the Periodic Table Books. There are just too many references to publish in this book but you can read and research them all at MagicalPTElements.com/MAUPT or /MDAPT or /MW1PT or /MW2PT or /MAKPT. There, you can also access book related activity sheets, coloring pages, games and even Elemental tee-shirts to help make the learning process more fun. You can see a few fun video shorts there, too.

Get ready made trading cards, lapel pins, tee shirts and more based on this book from Sybrina Publishing's No Metal No Magic Collection at Zazzle - **http://bit.ly/3km64Wg**

Would you like a 24" x 36" poster of the Elemental-Themed Periodic Table in this book? The best place to get it is at **https://bit.ly/49QMxBT** They have the sharpest images of any other poster printer around.

The Magical Elements of the Periodic Table books came into existence because of my Blue Unicorn—Journey To Osm books. If it weren't for their magical powers, based on the properties of the metals of their horns and hooves, I would have never come up with the idea to relate magical creatures to the periodic table. There's a metal horn unicorn story for every age group and they are all available at Sybrina's Unicorn Book Store on at Sybrina.com.

If you enjoyed this book
please leave a nice review
at your favorite online book site.

No Metal, No Magic

Song Lyrics

No metal, no Magic

No metal, no Magic

I can think of nothing more tragic

Than to have no metal or no magic

Metal makes everything magical.

Just ask a unicorn. . .

Preferably, one with a metal horn.

They'd say No metal, No magic.

Metal makes everything techno magical.

No metal, No magic

for two-leggers or unicorns.

No metal, No magic

Metal makes everything techno magical.

No metal, No magic

It's techno magical.

No metal, No magic

It might be very hard to believe but with

No metal, No magic

There'd be no technology.

No metal, No magic

No metal, No magic

Listen to this song at https://youtu.be/tcB8KDWAd8w

Watch the book trailer at https://youtu.be/NIX9fE7GJRI

Get These Trading Cards Sets Featuring The Magical Elementals Representing The Periodic Table Elements

at https://bit.ly/40oEUBr

Unicorns, Dragons, Wizards or Knights?

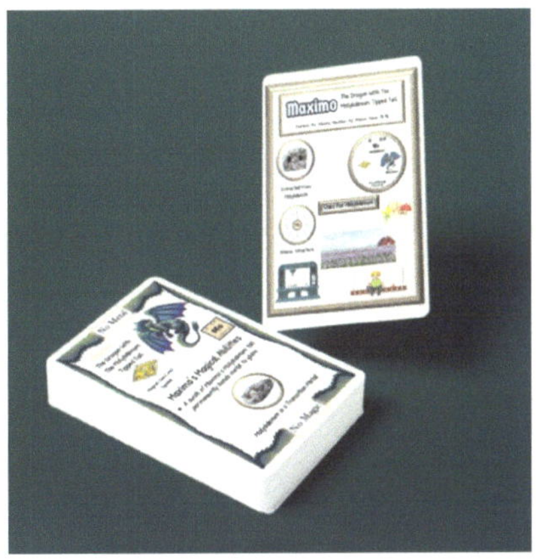

Collect Them All

These magical elementals are ready to help make learning the periodic table more fun. Get all books and related activities today.

Learn More About all of the Periodic Table Elementals.
Get all of the "No Metal No Magic" Books Featuring Individual Elements at MagicalPTElements.com

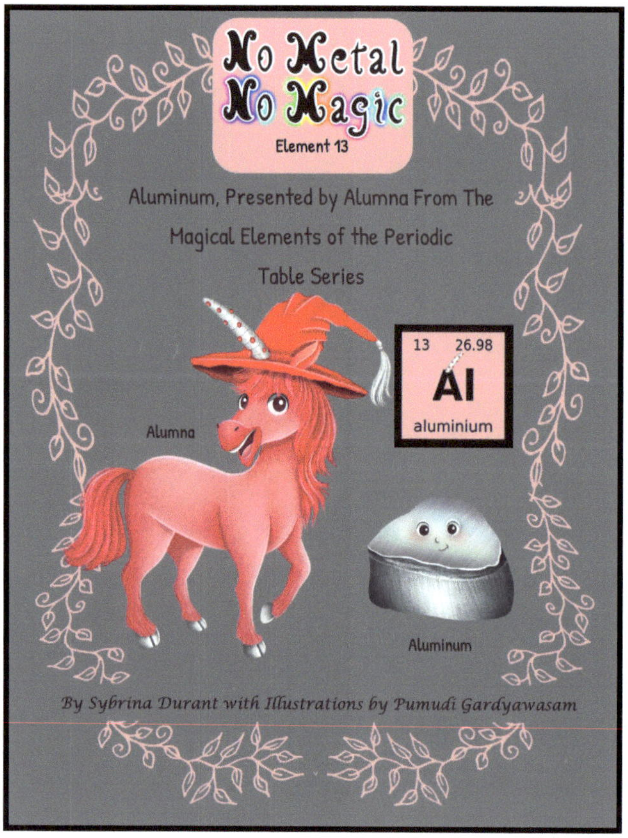

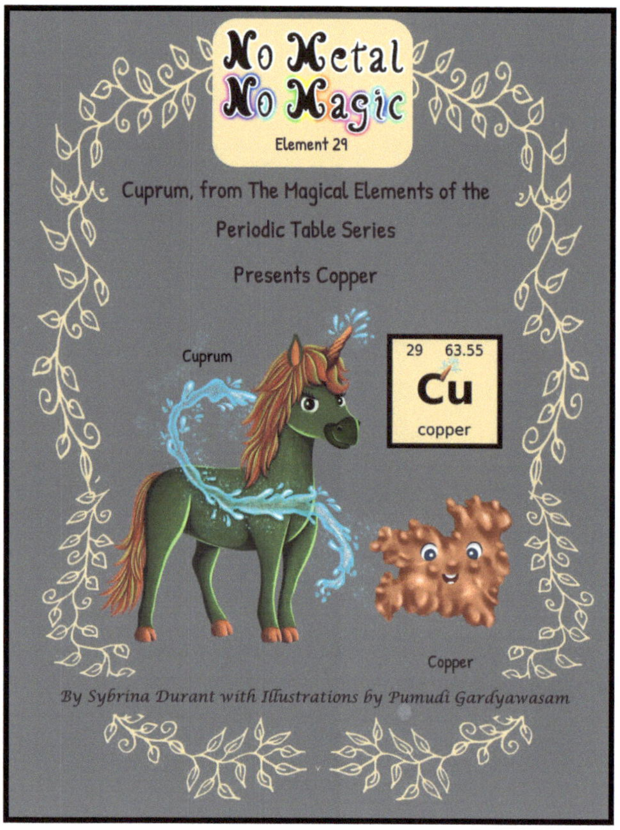

Don't Forget To Get A Tee Shirt

Featuring Your Favorite of the

118 Elements from the Periodic Table

Available in adult and kid sizes in many colors.

https://amzn.to/47NVZWN

Also Available From Sybrina Publishing

Magical Elemental-Themed Periodic Table

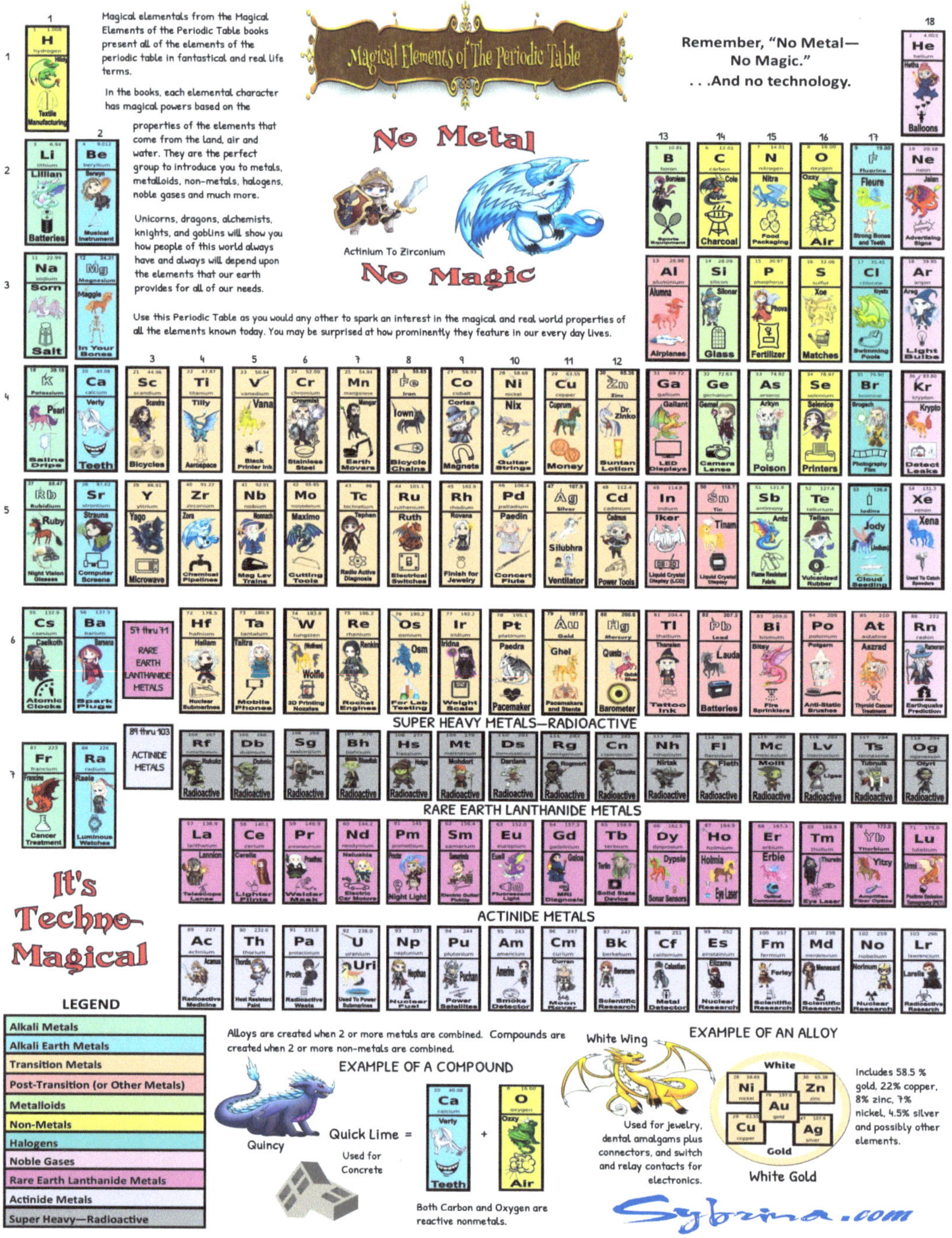

Would you like a 24" x 36" poster of the Magical Elemental-Themed Periodic Table from The Magical Elements of the Periodic Table books? The best place to get a high quality poster size print is at https://bit.ly/49QMxBT It will look great on a classroom or kid's room wall.

Get These Fun Elemental Periodic Table Activities at

MagicalPTElements.

Unicorn Periodic Table Bingo—Comes with 32 unique Bingo cards. Magical Elementals Bingo comes with 36.

Magical Elemental Game Cards—Makes great prizes. Fun to trade, too.

Also browse activities at

https://www.magicalPTelements.com

for all kinds of printable downloads to make learning fun.

Printable Magical Elemental Activity Downloads

Fun Way For Students To Learn The Elements Of The Periodic Table

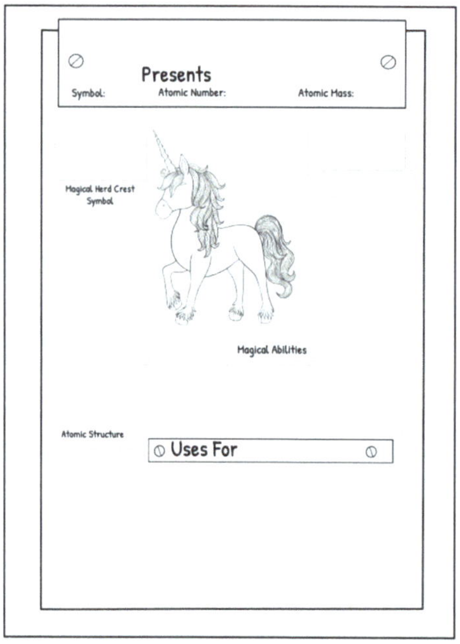

Blank Unicorn Element Card

Sample Unicorn Element Card

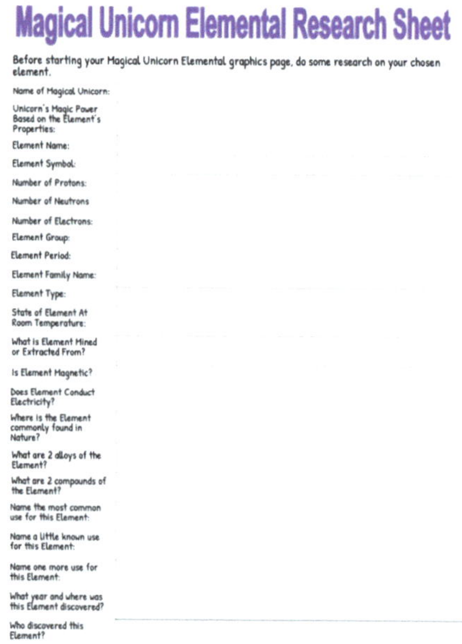

Magical Unicorn Elemental Research Sheet

Before starting your Magical Unicorn Elemental graphics page, do some research on your chosen element.

Name of Magical Unicorn:
Unicorn's Magic Power Based on the Element's Properties:
Element Name:
Element Symbol:
Number of Protons:
Number of Neutrons
Number of Electrons:
Element Group:
Element Period:
Element Family Name:
Element Type:
State of Element At Room Temperature:
What is Element Mined or Extracted From?
Is Element Magnetic?
Does Element Conduct Electricity?
Where is the Element commonly found in Nature?
What are 2 alloys of the Element?
What are 2 compounds of the Element?
Name the most common use for this Element:
Name a little known use for this Element:
Name one more use for this Element:
What year and where was this Element discovered?
Who discovered this Element?

Blank Research Sheet

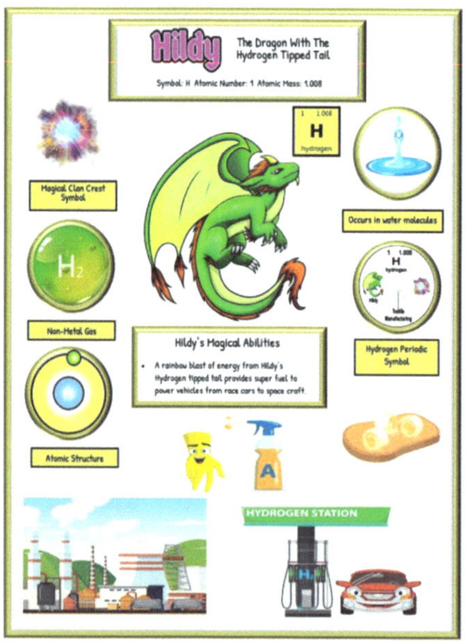

Sample Dragon Element Card

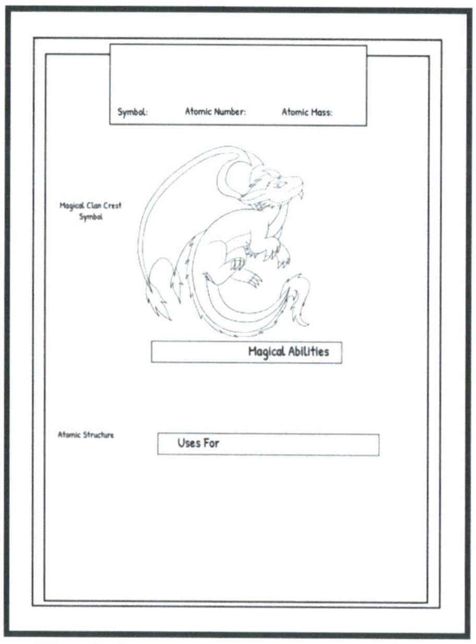

Blank Dragon Element Card

Magical Dragon Elemental Research Sheet

Before starting your Magical Dragon Elemental graphics page, do some research on your chosen element.

Name of Magical Dragon:
Dragon's Magic Power Based on the Element's Properties:
Magical Clan Crest Symbol:
Element Name:
Element Symbol:
Atomic Number:
Atomic Mass:
What year and where was this Element discovered?
Who discovered this Element?
Element Group:
Element Period:
Element Family Name:
State of Element At Room Temperature:
What is Element Mined or Extracted From?
Is Element Magnetic?
Does Element Conduct Electricity?
Where is the Element commonly found in Nature?
What is 1 alloy of the Element? How used?
What is 1 compound of the Element? How used?
Name the most common use for this Element:
Name a little known use for this Element:
Name one more use for this Element:
Interesting and Fun Facts:

Blank Research Sheet

Using the sample Magical Elemental cards provided, have students select an element from the Periodic Table and a Magical Elemental Card Blank to create their own Magical Elemental Card. The blank and sample cards do not have to match.

You will receive a pdf containing either 26 unicorn or 26 dragon sample cards and blanks to be printed on 8 1/2 x 11 sized paper or card stock. The pdf also contains a Magical Elemental Research Sheet for the students to work on before creating their unique Periodic Table Elemental. They will also write a short paragraph describing their Unicorn or Dragon Elemental from that research.

Excerpt from Chapter 24 (Rag-Bag Magic) of the novel,

Journey To Osm—The Blue Unicorn's Tale

Back then, most places throughout MarBryn had wizards and sorcerers of some ability, or another. Most were trained in the ways of magical arts by the unicorns as part of their outreach program. Some two-leggers developed practical magical skills like making delicious feasts of tasty food appear out of thin air or purifying murky water around the land.

Others went through more extensive training to learn battle magic—like shooting powerful streams of energy from their swords.

Some were taught the art of holding the glow of the sun in magical globes, bringing light into the dark of night. These magical lights warded off the evil beings that were new and frightening products of dark magic.

With the rise of the sorcerer Magh, magical defense arts had become more important. The highest level of magical training involved sensing when others were in danger and learning to see into the future. Very few two-leggers ever reached that level because magic wasn't inherent in them the way it was for the unicorns. The metal of their horns and hooves were part of them as well as the very makeup of their blood, but two-leggers relied on learned magic via potions, charms, and incantations that required help from ingredients and forces more mystical in nature than any two-legger was ever born to be. Of course, controlling magic and projecting your intentions went far beyond merely following a recipe of sorts. It took being in touch with nature and the various elements to get the response a wizard desired. Much trial and error went into it, as well as faith and trust and the motives of the spell caster.

Magic was and is a practice that is never quite perfected even for the unicorns who must continue to hone and learn how to harness their powers. On rare occasion a wizard and unicorn had formed enough trust and a steadfast bond that prompted the unicorn to gift the wizard with a wand or staff embedded with the smallest sliver of metal from one of their hooves. This was rare but had happened and of course so had the desire for more power. Magh wasn't the first sorcerer with lust for more and throughout history there had been a handful of heinous acts against unicorns from those seeking their magic. Prior to Magh those wizards had failed to circumvent the protections nature had infused unicorn magic with so even after harvesting metal from their horns or hooves these sorcerers had gone mad trying to bend the will of nature and actually use their ill-gotten gains.

Magh, too, had gone mad or perhaps he'd already been so, but somehow he'd managed to harness the metal he harvested from his victims and continued to grow stronger rather than completely lose his mind like the others. How exactly remained a mystery to the unicorns and everyone else.

The wizards who'd honed their craft with the help and blessing of the unicorns tried to solve the riddle and even the best of them failed to uncover that secret. Some of MarBryn's natives took to magic naturally, while others struggled with the concept but no matter how adept they were. All of them fought valiantly against Magh's magic because with even a shred of knowledge they understood how the power shift would ultimately play out. They fought to the end but, in the end, only one sorcerer remained in MarBryn and now, Magh was in total control. But still, he was not satisfied. He wanted to control every living creature in the land.

www.ingramcontent.com/pod-product-compliance
Lightning Source LLC
LaVergne TN
LVHW071240210926
848486LV00007B/314